Advances in Distillation Retrofit

Nguyen Van Duc Long · Moonyong Lee

Advances in Distillation Retrofit

Springer

Nguyen Van Duc Long
School of Chemical Engineering
Yeungnam University
Gyeongsan
South Korea

Moonyong Lee
School of Chemical Engineering
Yeungnam University
Gyeongsan
South Korea

ISBN 978-981-10-5899-8 ISBN 978-981-10-5901-8 (eBook)
DOI 10.1007/978-981-10-5901-8

Library of Congress Control Number: 2017954336

Printed on acid-free paper

This Springer imprint is published by Springer Nature
The registered company is Springer Nature Singapore Pte Ltd.
The registered company address is: 152 Beach Road, #21-01/04 Gateway East, Singapore 189721, Singapore

Preface

As a thermal separation method, distillation is considered the most important and mature separation technology in the chemical process industry. While this unit operation has many advantages, one major drawback is its large energy requirement, which significantly influence the overall plant profitability. Therefore, the increasing cost of energy has forced industry to reduce its energy consumption. In addition, many processes are required to increase capacity due to an increase of demand. To accomplish these tasks, retrofit of distillation processes become an important issue. To carry out effectively retrofit, an innovation solution is the use of advanced process integration and process intensification techniques, which has been a rapid growth in technological and commercial opportunities. In particular, multi-effect distillation, heat pump assisted distillation are the best examples of proven process integration technology, whereas thermally coupled distillation sequence (TCDS), dividing wall column (DWC), and reactive distillation are the best examples of proven process intensification technology in distillation because they have significantly lower investment and operating costs while also reducing the equipment and carbon footprint.

How can one decide quickly what techniques and whether they are good process concepts for distillation retrofit? This field requires a book that will capture an accurate snapshot regarding the field and provide an insightful review to all the key techniques in the revolution in distillation retrofit. Considering that there are already books dealing with known and basic issue in distillation, this book aims to provide the readers with some issues related to retrofit of distillation using advanced distillation techniques, emphasizing the use of multi-effect distillation, heat pump assisted distillation, TCDS, DWC, reactive distillation, and innovative hybrid systems, which have not been comprehensively discussed in the published books. Rather than prepare a textbook on advanced distillation in the usual and traditional format, the authors considered writing a book that highlighted concepts and practical applications rather than theory.

This book presents a comprehensive review of contemporary advanced techniques employed for enhancing the distillation process with the purpose is bridge the gap between developers of advanced distillation design procedures and those ultimately use them. Thus this book is a source of information for undergraduate and postgraduate students of chemical engineering, practicing process designers and chemical engineers.

Several companies supported our work. In particular we would like to thank Samsung BP Chemicals, Samsung Cheil Industries, SKC, LG Chemical, and KOGAS.

Nguyen Van Duc Long
Moonyong Lee

List of Abbreviations and Greek Letters

Abbreviations

A	Area [ft^2]
ABDWC	Azeotropic bottom dividing wall column
BDWC	Bottom dividing wall column
BC	Bare cost [$]
BMC	Updated bare module cost [$]
C_2	Ethane
C_3	Propane
C_{5+}	Gasoline
CD	Coordinate decent methodology
CE	Cellulosic ethanol
CO_2	Carbon dioxide
COP	Coefficient of performance
C_{steam}	Cost of the steam [$]
C_{CW}	Cost of cooling water [$]
$C_{electricity}$	Cost of electricity [$]
$C_{refrigeration}$	Cost of refrigeration [$]
CGCC	Column grand composite curve
D	Diameter [ft]
DWC	Dividing wall column
DWPC	Dividing wall prefractionator configuration
E_o	Overall efficiency [%]
E_{MV}	Murphree stage efficiency [%]
ECMD	Enhanced Capacity Multiple Downcomer
ETDWC	Extractive top dividing wall column
EPC	Engineering, procurement and construction
F	Feed flow [kg/hr]
FEED	Front-end engineering design

FL	Liquid flow [kg/hr]
FV	Vapor flow [kg/hr]
FTCDC	Fully thermally coupled distillation column
h	Enthalpy [kJ/kg]
HP	High pressure
iC_4	Isobutane
K	Distribution coefficient
L	Length [ft]
LLE	Liquid-liquid equilibrium
LP	Low pressure
MD	Multiple Downcomer
MF	Module factor
MPF	Material and pressure factor
MVR	Mechanical vapor recompression
N	Number of trays
nBuOH	Normal butyl alcohol
NGL	Natural gas liquid
nC_4	Normal butane
NRTL	Nonrandom two-liquid
Op	Operating cost [$]
P	Total vapor pressure of the solution
p_i	Partial pressure of component i in the vapor
p_i'	Vapor pressure of pure component i
PSA	Pressure swing adsorption
PFMD	Parallel Flow Multiple Downcomer
q	Feed thermal condition
R	Reflux ratio
RDWC	Reactive dividing wall column
RSM	Response surface methodology
R^2	Coefficient of determination
S	Area of the heat exchanger [ft^2]
SSC	Side stream column
T	Temperature [°C]
TSA	Temperature swing adsorption
TAC	Total annual cost [$]
TCDS	Thermally coupled distillation sequence
TCEDS-SR	Thermally coupled extractive distillation sequences
TCRDS	Thermally coupled reactive distillation sequence
TDWC	Top dividing wall column
VC	Vapor compression
VLE	Vapor-liquid equilibrium
x_i	Composition of component i in the liquid phase
y_i	Composition of component i in the vapor phase

UF	Update factor
UNIQUAC	Universal quasi-chemical
ΔT	Temperature difference

Greek letters

γ	Liquid activity coefficient
μ_i	Chemical potential of component i
α	Relative volatility
λ	Latent heat [kJ/kg]

Contents

About the Authors

Nguyen Van Duc Long is currently a Foreign Assistant Professor at Yeungnam University. He received his PhD in the Department of Chemical Engineering at Yeungnam University in August, 2012. Subsequently, he joined Samsung Cheil Industries as a Senior Process Engineer. He has worked on many long-term industrial projects, which were collaborative efforts between his laboratory and industrial manufacturers or engineering companies. These industrial projects mainly focus on process systems engineering, process retrofit and debottlenecking technologies, process integration and intensification technologies, dividing wall column, thermally coupled distillation sequence, reactive distillation, extractive distillation, azeotropic distillation, pressure swing distillation, batch distillation, heat pump assisted distillation, self-heat recuperation technology, separation processes, reactive separation processes, sustainable chemical processes, exergy and energy efficiency, natural gas processing, petroleum refinery, carbon capture, storage and utilization, carbon dioxide emission reduction, gasification, biofuels and green energy, biochemicals, biomass conversion and biorefineries. He published 38 scientific articles in peer-reviewed journals, 3 review papers and 1 book chapter.

Moonyong Lee has been a professor at Yeungnam University, School of Chemical Engineering, since 1994. Before joining the university, he had worked as Senior Process Engineer in SK refinery and petrochemical company for 10 years. He has an MSc and a PhD in Chemical Engineering from KAIST. Since entering the university, he has worked on many industrial projects related to process modeling, design, and control. His current research areas are design, modeling, optimization, and control of various energy and chemical processes (especially distillation based separation processes). He is the author of over 170 publications in international journals and the recipient of the GoMyeungBok Scientific Award from ICASE (2006), Conventional Chemical Engineering Special Award from KIChE (2013), and Korean Government Minister Award for Engineering Industry Contribution (2016).

Chapter 1
Introduction

1.1 Introduction

Without distillation, there would be no modern living (Rosenzeig, 1997) because it is the most well-defined separation process that is used widely to produce the world's petroleum fuels, to treat most natural gas, and is a critical element in a host of processes making the chemicals and other products that world needs (Sorensen and Darton, 2006). The performance of these industries is strongly dependent on the engineering and engineers. One typical drawback of distillation processes is that they require the largest portion of energy in industrial separation technologies with an estimated 3% of the world's energy consumption (Hewitt et al., 1999). This can significantly influence the overall plant profitability. Therefore, development of a new type of column and/or performance enhancement of existing distillation processes for improved energy efficiency have been imperative issues associated with distillation (Olujic et al., 2009). In addition, saving energy in this process is also an important issue from an environmental standpoint or carbon dioxide (CO_2) mitigation because the huge amount of energy consumed in distillation has a large impact on greenhouse gas emissions (Matsuda et al., 2012).

A broad range from improving capacity, energy and separation efficiency using process modification and/or process integration, such as multi-effect distillation or heat pump assisted distillation and/or process intensification, such as reactive distillation, and/or innovative hybrid systems, such as configuration combining distillation and pervaporation, have created enormous concern. Basic knowledge of distillation is required to develop a new type of column and/or enhance the performance of existing distillation processes for improved separation and energy efficiency.

1.2 Distillation Fundamentals and Principles

The separation operation in distillation is based on intimate contact between the vapor and liquid phases using various kinds of devices, such as random or structured packings and plates or trays. Because of the difference in gravity between

N.V.D. Long, M. Lee, *Advances in Distillation Retrofit*, DOI 10.1007/978-981-10-5901-8_1

the vapor and liquid phases, liquid runs down the column, cascading from tray to tray, while vapor flows up the column, contacting liquid at each tray (Seader et al., 2008). Liquid reaching the bottom of the column is vaporized partially in a heated reboiler to provide boil-up, which is sent back up the column, while the remainder of the bottom liquid is withdrawn as the bottom product. Vapor reaching the top of the column is cooled and condensed to a liquid in the overhead condenser. A part of this liquid is returned to the column as reflux to provide liquid overflow, while the remainder of the overhead stream is withdrawn as distillate, or overhead product. In some cases, only a part of the vapor is condensed so that a vapor distillate can be withdrawn. Note that the vapor and liquid leaving an equilibrium stage are assumed to be in complete equilibrium with each other; thus, thermodynamic relations can be used to determine the temperature and relate the concentrations in the equilibrium stream at a given pressure (Seader et al., 2008).

The lighter (lower-boiling) components tend to concentrate in the vapor phase, whereas the heavier (higher-boiling) components tend toward the liquid phase (Seader et al., 2008). The result is a vapor phase that becomes richer in the lighter components as it passes up the column and a liquid phase that becomes richer in heavy components as it cascades downward. The overall separation achieved between the distillate and the bottom depends primarily on the relative volatilities of the components, the number of contacting trays, and the ratio of the liquid phase flow rate to the vapor-phase flow rate.

1.2.1 Vapor Pressure

The vapor pressure, which is an important property of liquids, is the pressure at which both liquid and vapor exist at a given temperature. At this point, dynamic equilibrium exists, in which vaporization and condensation take place at equal rates and the pressure in the vapor space remains constant (Theodore and Ricci, 2010). The pressure at equilibrium is equal to the vapor pressure of the liquid. The vapor pressure is related to the boiling temperature. In particular, liquids with a high vapor pressure boil at low temperatures while liquids with a low vapor pressure boil at high temperatures. The vapor pressure depends only on temperature (Fig. 1.1). In particular, the vapor pressure increases with increasing temperature.

Raoult's law states that the partial pressure of each component (p_i) in a solution is proportional to the mole fraction (x_i) of that component in the liquid mixture being studied with the "proportionality constant" being its vapor pressure ($p_i^{'}$) (Theodore and Ricci, 2010). Therefore, for component i in a mixture, Raoult's law can be expressed as

$$p_i = p_i^{'} x_i \tag{1.1}$$

where p_i is the partial pressure of component i in the vapor, $p_i^{'}$ is the vapor pressure of pure component i at the same temperature, and x_i is the mole fraction of

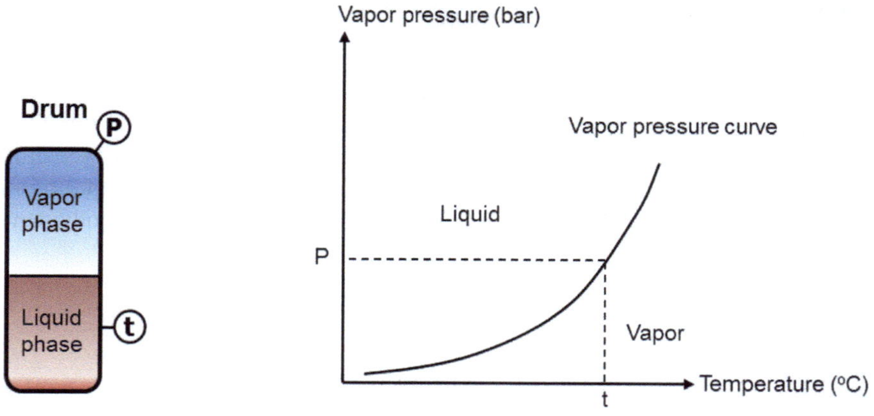

Fig. 1.1 Vapor pressure of a pure component

component i in the liquid. This expression may be applied to all components so that the total pressure, P, is given by the sum of all the partial pressures.

If the gas phase is ideal, Dalton's law applies

$$p_i = y_i P \tag{1.2}$$

where y_i is the mole fraction of component i in the vapor, and P is the total vapor pressure of the solution, which can be calculated as follows:

$$P = \sum_{i=1}^{N_C} p_i = \sum_{i=1}^{N_C} p_i' x_i \tag{1.3}$$

Therefore, Eq. (1.1) can then be written as

$$y_i = x_i \left(\frac{p_i'}{P} \right) \tag{1.4}$$

Raoult's law applications require vapor pressure information, which can be found in the literature (Green and Perry, 2008). However, there are two equations that can be used in lieu of the vapor pressure information, the Clapeyron equation and the Antoine equation (Theodore et al., 2009). The two constant (A, B) Clapeyron equation is given by

$$\ln p' = A - \frac{B}{T} \tag{1.5}$$

where p' and T are the vapor pressure and temperature, respectively. The three constant (A, B, C) Antoine equation is given by

$$\ln p' = A - \frac{B}{T + C} \tag{1.6}$$

1.2.2 Vapor-Liquid Equilibrium

The equilibrium principles between the vapor and liquid require understanding because the separation of chemical mixtures in distillation is based on the phase equilibrium behavior. They play an important role in designing and predicting the performance of distillation processes. Consider the vapor–liquid equilibrium of mixtures, where each component appears in both phases. At thermodynamic equilibrium,

$$P^V = P^L \tag{1.7}$$

$$T^V = T^L \tag{1.8}$$

$$\mu_i^V = \mu_i^L \tag{1.9}$$

where P, T and μ_i are the pressure, temperature, and chemical potential of component i, respectively.

Because the pressure and temperature fix the equilibrium vapor and liquid compositions, the experimental data is frequently presented in the form of tables of the vapor mole fraction y and liquid mole fraction, x, for one constituent over a range of temperatures, T, for a fixed pressure, P, or over a range of pressures for a fixed temperature (Seader et al., 2008).

The distribution of a component in two phases is represented by the distribution coefficient, K, as follows:

$$K = \frac{y_i}{x_i} \tag{1.10}$$

where y_i and x_i are the composition of component i in the vapor phase and liquid phase, respectively. A component with a high K-value tends to concentrate in the vapor, while it tends to concentrate in the liquid with a low K-value. The K-value is a function of temperature and pressure. The K values are used widely in distillation calculations, and the ratio of the K values of two components is defined as the relative volatility,

$$\alpha_{ij} = \frac{K_i}{K_j} = \frac{y_i/x_i}{y_j/x_j} \tag{1.11}$$

The relative volatility is a measure of the ease of separation, in particular, the higher it is, the easier the separation. For binary mixtures:

$$\alpha_{12} = \frac{K_1}{K_2} = \frac{y_1/x_1}{y_2/x_2} = \frac{y_1(1-x_1)}{x_1(1-y_1)} \tag{1.12}$$

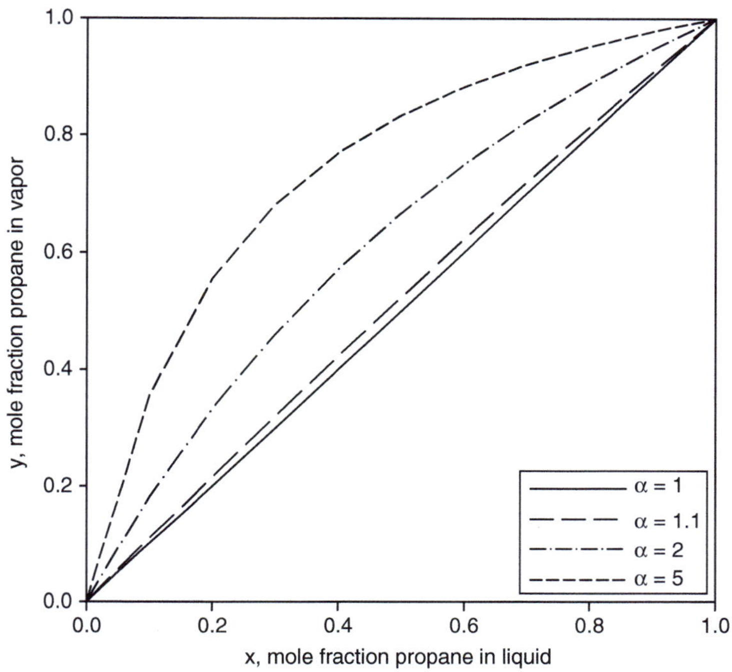

Fig. 1.2 *x–y* diagram with various relative volatilities

Rearranging this equation, results in

$$y_1 = \frac{\alpha_{12}x_1}{1 + (\alpha_{12} - 1)x_1} \tag{1.13}$$

Equation (1.13) expresses the more volatile component mole fraction in the vapor as a function of the mole fraction of that component in the liquid and the relative volatility (Kister, 1992), which can be plotted in Fig. 1.2 as an *x–y* diagram.

1.2.3 Dew Point and Bubble Point

To estimate the stage, condenser, and reboiler temperatures, set procedures are required for calculating the dew and bubble points (Towler and Sinnott, 2013). At the bubble-point temperature, the total vapor pressure exerted by the mixture becomes equal to the confining drum pressure, and it follows that $\sum y_i = \sum Kx_i = 1.0$ in the bubble formed (Seader et al., 2008). At the dew-point temperature, the relationship, $\sum x_i = \sum \frac{y_i}{K_i} = 1.0$, must be satisfied. A calculation of these temperatures must be found by iteration at a given system pressure.

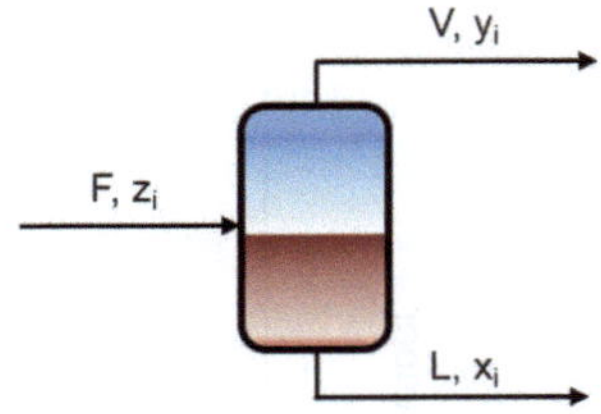

Fig. 1.3 Flash distillation

1.2.4 Equilibrium Flash Calculations

Suppose the temperature, pressure, and overall composition of the inlet stream shown in Fig. 1.3 is known. The mole fractions of component i in the inlet, vapor, and liquid streams are z_i, x_i and y_i, respectively, with $\sum y_i = 1.0$ and $\sum x_i = 1.0$. Subtracting these two equations results in:

$$\sum y_i - \sum x_i = 0 \tag{1.14}$$

When the vapor and liquid are in equilibrium, the relationship is $y_i = K_i x_i$. Equation (1.14) can be rearranged to:

$$\sum K_i x_i - \sum x_i = 0 \quad \text{or} \quad \sum (K_i - 1) x_i = 0. \tag{1.15}$$

The mole balance for each component over the phase separator (Finlayson, 2006) is

$$F = V + L \quad \text{and} \quad F z_i = V y_i + L x_i. \tag{1.16}$$

Dividing by F and defining v as the fraction of the feed that is vapor gives the following:

$$z_i = y_i v' + x_i (1 - v) \quad \text{with} \quad v = \frac{V}{F}. \tag{1.17}$$

Using the equilibrium expression again, Eq. (1.17) can be written as

$$z_i = K_i x_i v + x_i (1 - v) = (K_i - 1) x_i v + x_i. \tag{1.18}$$

Solving for the mole fractions in the liquid gives

$$x_i = \frac{z_i}{1 + (K_i - 1) v}. \tag{1.19}$$

Placing this expression into Eq. (1.15) results in the final equation.

$$\sum \frac{(K_i - 1)z_i}{1 + (K_i - 1)v} = 0 \tag{1.20}$$

This is called the Rachford–Rice equation (Finlayson, 2006). If the K values and inlet composition z_i are known, this is a nonlinear equation to solve for v. Once the value of v is known, the value of the liquid composition x_i, vapor composition y_i, vapor flow rate, and liquid flow rate can be calculated.

1.3 Distillation Design

1.3.1 Shortcut Methods

Shortcut methods can be used to estimate the stage and energy requirements without the aid of computers. They generally depend on the assumption of constant relative volatility, and should not be used for severely non-ideal systems (Towler and Sinnott, 2013).

The minimum number of theoretical stages at the total reflux (N_m) was estimated using the Fenske equation (Fenske, 1932).

$$N_{\min} = \frac{\ln\left[\left(\frac{x_{D,LK}}{x_{D,HK}}\right) / \left(\frac{x_{B,LK}}{x_{B,HK}}\right)\right]}{\ln \alpha_{LK-HK}} \tag{1.21}$$

where $\alpha_{LK\text{–}HK}$ is the average relative volatility of the light key with respect to the heavy key; and x_{LK} and x_{HK} are the light and heavy key concentration, respectively.

The Underwood equations calculating the minimum reflux ratio in a multicomponent system is used widely as follows:

$$\sum \frac{\alpha_i x_{D,i}}{\alpha_i - \theta} = R_{\min} + 1 \tag{1.22}$$

where θ is the root of the equation,

$$\sum \frac{\alpha_i x_{F,i}}{\alpha_i - \theta} = 1 - q \tag{1.23}$$

where q is the feed thermal condition with quantity is the liquid fraction of the feed.

Many reflux-stages relationships have been proposed with the most popular method being the Gilliland correlation. This plot correlating the reflux and stages can be used to estimate the number of stages, N, after N_{min}, minimum reflux ratio ($R_{\min}$) and reflux ratio (R) are determined using the Fenske and Underwood equations. The procedure uses a diagram plot $(R - R_{\min})/(R + 1)$ on the x-axis and

$(N - N_{min})/(N + 1)$ on the y-axis (Kister, 1992). One enters the diagram with the abscissa value, which is known, and then it reads the ordinate of the corresponding point on the Gilliland curve (Kiss, 2013). The only unknown of the ordinate is the total number of stages (N).

To estimate the feed stage location, the most popular shortcut relationships are the Fenske equation and the Kirkbridge equation. The Kirkbridge equation is expressed as

$$\frac{N_R}{N_S} = \left[\frac{z_{HK}}{z_{LK}}\left(\frac{x_{B,LK}}{x_{D,HK}}\right)^2 \frac{B}{D}\right]^{0.206}. \tag{1.24}$$

1.3.2 Rigorous Methods

Although shortcut methods can be performed simply and fast by hand, they have inferior accuracy. With the rapid development of computer technology, rigorous calculations describing a column as a group of equations and solving these equations to calculate the operating conditions of the column are the primary tools for the design and optimization in modern distillation practice. In particular, rigorous methods are used to solve the MESH equations, which stand for (Kister, 1992):

- Material or flow rate balance equations, both component and total,
- Equilibrium equations, including the bubble-point and dew-point equations,
- Summation or stoichiometric equations or composition constraints, and
- Heat or enthalpy or energy balance equations.

Normally the inside-out method is used to calculate most columns because of its robustness and its ability to solve a wide variety of columns. On the other hand, it is incapable of handling reactions in the column, which can be changed to rate-based methods.

1.4 Distillation Simulation

The analysis, conceptual design, optimization, operation, and control of distillation processes can be done using process simulators, such as Aspen Plus, Aspen Hysys, ProII, which are now available and accessed.

1.4.1 Azeotrope Search

Many mixtures of chemical components form azeotropes, which have vapor and liquid phases with identical compositions. This phenomenon affects the distillation design and simulation in terms of the feasibility, effort and time. These mixtures can be checked easily as to whether an azeotropic mixture exists or not using Aspen Plus.

1.4.2 Physical Property Methods

Because there are no solutions in reality, Raoult's law will provide highly inaccurate results. This is encountered frequently in distillation, namely, where the liquid phase is not an ideal solution but the pressure is low enough so that the vapor phase behaves as an ideal gas (Theodore and Ricci, 2010). In this case, the deviations from ideality are localized in the liquid and treatment is possible by quantitatively considering the deviations from Raoult's law. These deviations are taken into consideration by incorporating a correction factor, γ, into Raoult's law. The purpose of γ, which is defined as the activity coefficient, is to account for the departure of the liquid phase from ideal solution behavior. This is introduced into the Raoult's law equation as follows:

$$y_i P = p_i = \gamma_i x_i p_i' \tag{1.25}$$

The computation of the liquid activity coefficient, γ, can be done using several models, such as nonrandom two-liquid (NRTL), van Laar, Wilson, and universal quasi-chemical (UNIQUAC). Therefore, an extremely important issue in distillation simulations is the selection of an appropriate physical property method that will describe the phase equilibrium of the chemical system accurately (Kiss, 2013). Missing or inadequate physical properties can undermine the accuracy of a model or even prevent one from performing the simulation.

One of the most popular property methods is the NRTL model, which calculates the liquid activity coefficients for the following property methods: NRTL, NRTL-2, NRTL-HOC, NRTL-NTH, and NRTL-RK. This is recommended for highly non-ideal chemical systems, and can be used for vapor-liquid equilibrium (VLE) and liquid-liquid equilibrium (LLE) applications (Aspen Technology, 2009). The equation for the NRTL model is as follows:

$$\ln \gamma_i = \frac{\sum_j x_j \tau_{ji} G_{ji}}{\sum_k x_k G_{ki}} + \sum_j \frac{x_j G_{ij}}{\sum_k x_k G_{kj}} \left(\tau_{ji} - \frac{\sum_m x_m \tau_{mj} G_{mj}}{\sum_k x_k G_{kj}} \right) \tag{1.26}$$

where

$$G_{ij} = \exp(-\alpha_j \tau_{ji}) \tag{1.27}$$

$$\tau_{ji} = a_{ij} + \frac{b_{ij}}{T} + e_{ij} \ln T + f_{ij} T \tag{1.28}$$

$$\alpha_{ji} = c_{ij} + d_{ij}(T - 273.15\text{K}) \tag{1.29}$$

$$\tau_{ii} = 0 \tag{1.30}$$

$$G_{ii} = 1 \tag{1.31}$$

The binary parameters, a_{ij}, b_{ij}, e_{ij}, and f_{ij} are unsymmetrical. The Aspen Physical Property System has a large number of built-in binary parameters for the NRTL model. The binary parameters have been regressed using the VLE or LLE data from the Dortmund Databank (Luyben and Chien, 2010; Dortmund Databank (DDB). http://www.ddbst.de/).

1.4.3 Binary Parameters

Normally, simulators, such as Aspen Plus, Aspen Hysys or ProII, have binary parameters for popular mixtures. In addition, the DECHEMA Chemistry Data Series contains a large number of binary parameters, which have been obtained by regression using VLE or LLE experimental data (Luyben and Chien, 2010; Gmehling and Onken, 1977; Sørensen and Arlt, 1979). In Aspen Plus, the binary parameters can be obtained by data regression using the experimental data of *Txy* at a specific pressure.

1.5 Distillation Process Retrofit/Revamp and Debottlenecking

There is a significant need for continuous distillation process modification and improvement by target key goals, such as conserving natural resources, reducing energy required and/or operating cost, improving conversion and/or selectivity of reactions, enhancing productivity, using feed of different quality and/or alternative feed, achieving the new specifications of products, enhancing the control of the process, improving the safety, reliability and flexibility of the process and mitigating the discharge of ecologically harmful materials (El-Halwagi, 2006; Rangaiah, 2016). Besides, new and better technologies and separation processes as well as simulation and optimization methodologies are being continually developed and improved though research and industrial implementation (Rangaiah, 2016). These motivate researchers and practitioners more active to find suitable and advanced solutions to improve the existing plants through retrofit projects.

Process retrofit and revamp refers to making suitable changes and/or additions to existing process configuration and equipment (Rangaiah, 2016). According to Rangaiah (2016), retrofit can be used to mean adding to or replacing the whole or part of one type of equipment with a better alternative, and revamp for reorganizing the process involving several process steps (thus different equipment types).

Currently, retrofit and revamp are often used synonymously in the chemical engineering literature although some practitioners use retrofit for smaller projects (that is, investment) and revamp for bigger projects.

The most straightforward designs are grass-roots designs because they allow the most freedom to choose the design options and size of the equipment, whereas the design of a retrofit must attempt to work within the constraints of existing equipment (Smith, 2005). In other words, retrofits of distillation systems normally focus on the more efficient reuse of existing equipment to increase profit. Note that the retrofit of distillation columns is carried out more often than the installation of new equipment because distillation is an energy-intensive process that requires considerable capital investment (Gadalla et al., 2003). Retrofit projects comprised 70–80% of the capital investment projects in the processing industry at the end of the 1980s (Liu and Jobson, 2004).

In an existing distillation process with a fixed feed composition, there will be a maximum feed flow rate that can be separated to meet certain fixed product specifications (Liu and Jobson, 2004). Nevertheless, an increase in production capacity can cause bottleneck problems. The question is how engineers can identify opportunities or realize and/or remove these bottlenecks to achieve these goals methodically and efficiently from the complexities of the process. The debottlenecking of an existing facility is an attractive method due to the minimal cost required, particularly when compared with the construction of an additional facility (Schneider, 1997). This is possible because of the spare size/capacity available in many items of the existing equipment because of design margins used at the time of their design and fabrication (Rangaiah, 2016). Process debottlenecking can be achieved through retrofit the process so that it can be operated with higher production rate. Thus it can be considered as one case of process retrofit.

There are five main stages in the industrial retrofit projects including (1) establishing objectives and basis for retrofit, (2) conceptual process design, (3) front-end engineering design (FEED), (4) engineering, procurement and construction (EPC), and (5) commissioning/startup and completion (Rangaiah, 2016). Among these stages, the conceptual distillation process retrofit and debottlenecking designs rely on the creativity, solid technological knowledge, and broad experience of the design engineer (Skiborowski et al., 2014). Therefore, this has been referred to as rather an art than a science (Westerberg, 2004).

An innovative solution to solving retrofit projects is the use of advanced process integration and process intensification techniques. These projects within a distillation process generally cover a wide range of modifications, ranging from simple operating condition modifications or the replacement of column internals by high efficiency/capacity tray or packing to large modifications of the column configuration (Long and Lee, 2013) by heat integration or using intensified distillation columns, such as reactive distillation, dividing wall column, partial enlargement of the shell diameter and height, and/or modifications of auxiliary equipment. In any case, maximizing the use of existing equipment while, at the same time, minimizing the investment in new hardware is the key to a successful retrofit.

1.5.1 Process Integration

Because the process operates with multiple units and streams, there are a large number of process alternatives to accomplish the same goal, which can be eliminated by experienced designers. Normally, the chosen alternative is more attractive when the heating and cooling requirements can take place in the same heat exchanger to minimize the heating and cooling loads for that process. This can be referred to as process integration, which is defined as a holistic approach to design and operation that emphasizes the unity of the process (El-Halwagi, 2006). Furthermore, the increasing energy costs has accelerated the efforts of companies and universities to look for ways of improving energy efficiency. Process integration provides an excellent framework for benchmarking process performance, characterizing the root causes of problems limiting their performance, identifying opportunities, methodically generating efficient strategies that achieve the desired performance targets, and inventing innovative solutions (El-Halwagi, 2006). Therefore, process integration techniques are now widely used for plant optimization – both the optimization of new designs and improvements in existing facilities (Rossiter, 2003).

Process integration can be classified into two types, which are energy integration or heat integration and mass integration. Heat integration is a systematic methodology that provides a fundamental understanding of energy utilization within the process and employs this understanding in identifying the energy targets and optimizing heat-recovery and energy-utility systems (El-Halwagi, 2006). Mass integration is a systematic methodology that provides a fundamental understanding of the global flow of mass within the process and employs this understanding in identifying performance targets and optimizing the allocation, separation, and generation of streams and species (Dunn and El-Halwagi, 2003). Considerable effort in retrofit area has been directed toward increasing the heat recovery in chemical processes through heat integration (Long and Lee, 2014).

Some popular heat integration methods for enhancing the distillation processes are utilizing the heat from the top vapor stream or bottom product to preheat the feed or multi-effect distillation or heat pump assisted distillation mentioned in the following chapters. They have been applied successfully to reduce the energy requirements of processes, where all the units are configured with little or no integration. In this study, it is considered a critical element in retrofit designing for cost-effective and sustainable processes in general processes and for distillation processes in particular.

Note that when the process considered is relatively simple, heat exchanger network can be designed by the engineering insight to get the optimal or near-optimal heat integration solution. However, when the process has many hot and cold streams needed to exchange the heat using a complex heat exchanger network, the pinch technology can be normally employed. This can be conveniently accomplished using Aspen Energy Analyzer.

1.5.2 Process Intensification

The definition of process intensification can be interpreted usefully as being a process development involving the innovative design of multifunctional equipment, which leads to capital reduction (smaller, cheaper), safety (reduced volume), higher energy efficiency, environment impact, and improved raw material yields (Reay et al., 2013). Currently, process intensification is considered the main trend for improving the process performance (Rong and Turunen, 2006; Long et al., 2016). It is utilized widely in unit operations of heat transfer, reaction, separation, and mixing, which results in plant compactness, cleanliness, and energy efficiency. In particular, to intensify the separation process, there are two areas (Stankiewicz and Moulijn, 2000) – process-intensifying equipment, such as Higee referring rotating packed bed (RPB) (Ramshaw and Mallinson, 1981), and process-intensifying methods, such as hybrid process (Buchaly et al., 2007), dividing wall column (DWC) (Wright, 1949), integration of reaction and separation (Schmidt-Traub and Górak, 2006), and techniques using alternative energy sources, including microwave, centrifugal field and electric fields (Reay et al., 2013).

Among methods applying process intensification to enhance the performance of separation processes in general and distillation processes in particular, the thermal coupling technique provides a new mechanism for improving the traditional simple distillation column configuration through the principle of process intensification (Errico et al., 2008). This occurs by eliminating either the condenser or reboiler or both in the traditional simple column configuration. Thermal coupling can provide simultaneous mass and heat transfer between the different tasks in a separation sequence. Therefore, it provides a unique approach for intensifying the traditional simple distillation column configuration, which results in a decrease in the number of heat exchangers, improving the separation efficiency, and saving both energy and capital costs. A dividing wall column is the best example of distillation intensification, which have attracted considerable attention for consideration as a standard column for building a new plant as well as for improving the existing distillation columns.

Furthermore, over the years, the focus of the chemical and process industry has shifted towards the development and application of integrated processes combining the mechanism of reaction and separation in a single unit (Buchaly et al., 2007). This trend is motivated by benefits, such as a decrease in equipment and plant size and improvements of the process efficiency, and hence, better process economy. Reactive distillation is an important example of a reactive separation process.

In addition, several non-ideal aqueous–organic mixtures are formed, which tend to form azeotropes. This can be overcome using membrane separations, such as pervaporation and vapor permeation because they are very selective and not limited by vapor–liquid equilibrium (Buchaly et al., 2007). Consequently, a hybrid process consisting of membrane-assisted reactive distillation contributes to sustainable process improvement due to the synergy effects and allows a reduction of investment and operational costs.

Because process intensification can reduce the investment cost, inventory and improve heat management/energy utilization, it is expected to have wide ranging applicability, ranging from petrochemicals and bulk chemicals (Harmsen, 2010), fine chemical and pharmaceuticals industries (Reay et al., 2013) to biofuels and biochemicals (Nhien et al., 2017), carbon capture (Joel et al., 2014) and offshore processing (Lee et al., 2012). Especially, owing to the large volume production of petrochemicals and bulk chemicals, reducing energy consumption and environmental impact are significant motivations of technology innovation.

1.5.3 Combination of Heat Integration and Process Intensification

Heat integration and process intensification have been proven to be promising techniques in both academia and industry. Thus, combination of them is expected to bring a significant improvement. However, it causes the problem more complicated and requires a simple and efficient systematic methodology. Figure 1.4 shows a systematic framework combining heat integration and process intensification techniques to improve the performance of the chemical processes. This procedure essentially has several steps. First, the existing conventional chemical process or the base case was simulated, analyzed and evaluated. Next, heat integration techniques was studied for quick feasibility evaluation. In this step, the streams needed to be condensed and/or cooled and having high temperature can be used to exchange heat with the low-temperature streams that need to be heated and/or vaporized. Furthermore, when there is a small difference in temperature between the top and bottom of the column and the heat load is high, a heat pump system can be an

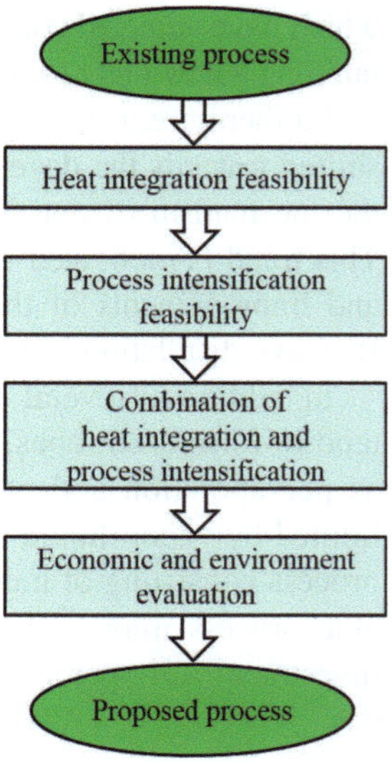

Fig. 1.4 Comprehensive procedure for improving the existing distillation process

attractive option for the recovery of heat released from the condenser to be used to boil the liquid in the reboiler. The process is then assessed to determine the feasibility of implementing process intensification methodologies. The DWC that is the most practical outstanding example of process intensification in distillation can be considered to improve energy and cost efficiency. Subsequently, both heat integration and process intensification techniques were applied to maximize the process performance. The promising process is finally proposed, which requires minimum operating cost, and has minimum total annual CO_2 emissions.

1.6 Concluding Remarks

The basis of vapor-liquid phase equilibrium as well as design methods, simulation and distillation equipment have been reviewed, which provides an overall understanding for the further design and improvement of a distillation column. This chapter also briefly defined process integration and process intensification, which are attractive innovative solutions for distillation process retrofit and debottlenecking.

References

Aspen Technology (2009) Aspen physical property system – physical property models, Version number V7.1

Buchaly C, Kreis P, Górak A (2007) Hybrid separation processes—combination of reactive distillation with membrane separation. Chem Eng Process 46:790–799

Dortmund Databank (DDB). http://www.ddbst.de/

Dunn RF, El-Halwagi MM (2003) Process integration technology review: background and applications in the chemical process industry. J Chem Tech Biotech. 78:1011–1121

El-Halwagi MM (2006) Process integration. In: Stephanopoulos G, Pistikopoulos E (eds) Process systems engineering, Elsevier, Amsterdam

Errico M, Rong BG, Tola G et al (2008) Process intensification for the retrofit of a multicomponent distillation plant-an industrial case study. Ind Eng Chem Res 47:1975–1980

Fenske MR (1932) Fractionation of straight-run Pennsylvania gasoline. Ind Eng Chem. 24:482–485

Finlayson BA (2006) Introduction to chemical engineering computing. Wiley, Hoboken

Green D, Perry R (eds) (2008) Perry's chemical engineers' handbook, 8th edn. McGraw-Hill, New York

Gadalla M, Jobson M, Smith R (2003) Shortcut models for retrofit design of distillation columns. Chem Eng Res Des 81:971–986

Gmehling J, Onken U (1977) Vapor–liquid equilibrium data collection. In: Behrens D, Eckermann R (eds) DECHEMA Chemistry Data Series. DECHEMA Publisher, Frankfurt

Harmsen J (2010) Process intensification in the petrochemicals industry: Drivers and hurdles for commercial implementation. Chem Eng Process 49:70–73

G. Hewitt, J. Quarini, M. Morrell (1999) More efficiency distillation, Chem Eng. (October)

Joel AS, Wang M, Ramshaw C, Oko E (2014) Process analysis of intensified absorber for post-combustion CO_2 capture through modelling and simulation. Int J Greenhouse Gas Control 21:91–100

Kiss AA (2013) Advanced distillation technologies: design, control and applications, 1st edn. Wiley, West Sussex

Kister HZ (1992) Distillation design. McGraw-Hill, New York
Lee SG, Long NVD, Lee MY (2012) Design and optimization of natural gas liquefaction and recovery processes for offshore floating liquid natural gas plants. Ind Eng Chem Res 51:10021–10030
Liu ZY, Jobson M (2004) Retrofit design for increasing the processing capacity of distillation columns: 1. A hydraulic performance indicator. Chem Eng Res Des. 82:3–9
Long NVD, Lee MY (2013) Debottlenecking the retrofitted thermally coupled distillation sequence. Ind Eng Chem Res. 52:12635–45
Long NVD, Lee MY (2014) Review of retrofitting distillation columns using thermally coupled distillation sequences and dividing wall columns to improve energy efficiency. J Chem Eng Japan 47:87–108
Long NVD, Minh LQ, Ahmad F, Luis P, Lee MY (2016) Intensified distillation-based separation processes: recent developments and perspectives. Chem. Eng. Technol. 39:2183–2195
Luyben WL, Chien IL (2010) Design and control of distillation systems for separating azeotropes, 1st edn. Wiley-AIChE, pp. 217–244
Matsuda K, Iwakabe K, Nakaiwa M (2012) Recent advances in internally heat-integrated distillation columns (HIDiC) for sustainable development. J Chem Eng Japan 45:363–372
Nhien LC, Long NVD, Lee MY (2017) Novel heat–integrated and intensified biorefinery process for cellulosic ethanol production from lignocellulosic biomass. Energ Convers Manage. 141:367–377
Olujic Z, Jödecke M, Shilkin A et al (2009) Equipment improvement trends in distillation. Chem Eng Process. 48:1089–1104
Ramshaw C, Mallinson RH (1981) Mass transfer process, U.S. Patent 4,283,255, 11 August
Rangaiah GP (2016) Chemical process retrofitting and revamping, Wiley, West Sussex
Reay D, Ramshaw C, Harvey A (2013) Engineering for efficiency, sustainability and flexibility. 2nd edn. Elsevier, Oxford
Rong BG, Turunen I (2006) Process intensification for systematic synthesis of new distillation systems with less than N-1 columns. Comput Aid Chem Eng 21:1009–1014
Rosenzweig M (1997) Preface. In: Nalven GF (ed) Practical engineering perspectives – distillation and other industrial separations, AIChE, New York
Rossiter AP (2003) Succeeding in process integration, process industries expo user conference.
Schmidt-Traub H, Górak A (2006) Integrated reaction and separation operations: modelling and experimental validation. Springer, Verlag Berlin Heidelberg
Schneider DF (1997) Debottlenecking options and optimization 1–15. http://www.stratusengr.com/Articles/DebottleOptions.pdf
Seader JD, Siirola Jeffrey J, Barnicki Scott D (2008) Distillation, in Don W. Green, Robert H. Perry, Perry's chemical engineers' handbook, 8th edn. McGraw-Hill, New York
Skiborowski M, Harwardt A, Marquardt W (2014) Conceptual design of azeotropic distillation processes. In: Gorak A and Sorensen E (eds) Distillation: fundamentals and principles, 1st edn. Elsevier, New York, pp 225–270
Smith R (2005) Chemical process design and integration. Wiley, West Sussex
Sørensen JM, Arlt W, (1979) Liquid–liquid equilibrium data collection. In: Behrens D, Eckermann R (eds) DECHEMA Chemistry data series. DECHEMA Publisher, Frankfurt
Sorensen E, Darton R Preface in distillation & absorption 2006
Stankiewicz A, Moulijn JA (2000) Process intensification: transforming chemical engineering. Chem Eng Prog. 96:22–24
Theodore L, Ricci F, (2010) Mass transfer operations for the practicing engineer. Wiley, Hoboken
Theodore L, Ricci F, Vliet TV (2009) Thermodynamics for the practicing engineer. Wiley, Hoboken
Towler G, Sinnott R (2013) Chemical engineering design - principles, practice and economics of plant and process design, 2nd edn, Elsevier, Oxford
Westerberg AW (2004) A retrospective on design and process synthesis. Comp Chem Eng. 28:447–458
Wright RO (1949) Fractionation apparatus, U.S. Patent 2,471,134, 24 May

Chapter 2
Retrofit and Debottlenecking by Modifying Column Internal

2.1 Introduction

Endeavoring towards greater sustainability drives the process industries to search for opportunities to improve the efficiency and/or capacity of distillation columns and sequences. Numerous academics and practitioners are active in this field and utilize a range of approaches to provide theoretically sound conceptual, technology advanced solutions. Two main approaches have been adopted for the design of distillation system retrofits to increase the capacity and/or reduce the energy requirement (Liu and Jobson, 2004):

- replacing existing internals with high-capacity and/or high-efficiency internals; and
- revamping the process by improving the utilization of existing equipment and making relatively minor modifications, including adjusting the operating conditions, adding equipment, etc.

Replacing existing column internals not only promotes separation, but also governs the column hydraulic performance (Amminudin and Smith, 2001). The retrofit of an existing column, such as the replacement of existing trays or packings with effective alternatives, is a frequent option because the user companies strive to increase production with minimal investment. Normally, vacuum applications are dominated by structured packings, whereas trays are normally used in high-pressure applications (above 5 bar) (Olujić et al., 2009). In addition, random packings can find favorable applications in the complete operating range. To cope with the ever growing demand for capacity, all families of vapor–liquid contactors have been subjected to intensive research and development in recent years (Schultes, 2003; Kashani et al., 2005; Weiland et al., 2005; Jödecke et al., 2006; Penciak et al., 2006; Jödecke et al., 2007; Darakchiev and Semkov, 2008). One excellent review (Olujić et al., 2009) provided a comprehensive overview of the equipment trends in distillation covering current column internals. On the other

N.V.D. Long, M. Lee, *Advances in Distillation Retrofit*, DOI 10.1007/978-981-10-5901-8_2

hand, only a few papers have been concerned with finding adequate technical solutions related to the column internal that could be implemented in industrial practice in a cost-effective manner. Note that the column internal refers to vapor-liquid contactor and others, such as vapor and liquid distributors, liquid collectors, demister, draw trays, chimney trays, packing support trays. In this chapter, only column internal trays and packings are mentioned.

2.2 Types of Column Internal

In a distillation column, vapor and liquid are contacted counter-currently. This is generally done in vertical cylindrical column containing trays or packings (random and structured) or both. As there are several excellent books providing the details of trays and packings, such as *Distillation Design* (Kister, 1992), *Distillation Tray Fundamentals* (Lockett, 1986), *Packed Tower Design and Applications: Random and Structured Packings* (Strigle, 1994), *Distillation Equipment and Processes* (*Górak* and Olujić, 2014), there is no need to go over these topics in detail. The authors will only describe briefly these trays and packings and why they sometimes have low efficiency or low capacity, and then move to high capacity/efficiency trays or packings and their use in a retrofit.

2.2.1 Trays

Trays are relatively complex distillation internals from a design and hydraulic standpoint, which consist of a tray deck, downcomer, and outlet weir. In the operation of a typical cross-flow tray, liquid enters the tray at one side and flows cross the tray toward a downcomer at the opposite side of the tray (Bruckert, 1968). As the liquid flows across the tray, vapor passing through the perforations in the tray surface forms a froth on the tray surface, which allows mass transfer to take place. At the outlet from the tray, the froth spills over a weir subtending the tray into the downcomer. The downcomer is a device for transferring the liquid phase downward to the next successive tray but is sealed by such liquid against the vapor passage upward. In the downcomer, the vapor separates from the froth, leaving a pool of disengaged liquid in the downcomer below the level of the active tray surface. The disengaged liquid enters the next tray under the force of its hydrostatic head, independently of the forces that create the froth and move the froth across the tray.

The tray decks commonly used in refinery and petrochemical plants can be divided into three types: sieve, valve, and bubble-cap. Bubble-cap trays, which were patented by Cellier-Blumenthal in 1815, were the most popular type of trays before the 1960s, when they were replaced with new designs of sieve and valve trays that matched the performance of bubble-cap trays but were much cheaper to install (Kister, 1992). Currently, bubble-cap trays are used only for special applications. Sieve trays, which are perforated thin plates, are very good in fouling applications or when there are solids present because they are easy to lean. However, the turndown (the performance when operating below the designed flow rate) is relatively poor (Wankat, 2012). Valve trays are designed to have better turndown properties than sieve trays. Therefore, they are more flexible when the feed rate varies.

2.2.2 Packings

Although trayed columns provide stepwise contact between the vapor and liquid phases, packed columns provide continuous contact without full disengagement of the vapor and liquid between the top and bottom of the packed bed (Pilling and Holden, 2009). In a packed column, the liquid driven by gravity flows down the mass transfer zone comprising the random and/or structured packings in the form of a trickle film or falling droplets, while the vapor as the continuous phase flows upwards from the bottom to the top of the column (Mackowiak and Macckowiak, 2014). Packings are generally divided into random and structured packings, which can be made from different materials, such as metal, plastic, and ceramics. Figure 2.1 shows the history of the development of characteristic random packings of different generations. Regarding the structured packings they can be classified into sheet, gauze, and grid packings (Spiegel and Duss, 2014). The main advantage of structured packings is their relatively low pressure drop per theoretical plate or equilibrium stage, which makes them the most effective vapor-liquid contactor for low-pressure, mainly vacuum applications (Olujić, 2014). Packing with a pressure drop that is about one-fifth that of the trays (Lockett, 1986) is often the preferred choice in cases where a pressure drop is the overriding consideration, such as vacuum distillation or where vapor recompression (heat pumping) is used.

The two features that should be maximized in packed beds are the open area, which is the average % of the cross-sectional area of the column not blocked by the packing, and hence available for the flow of vapor and liquid, and the wetted surface area, which is the number of square feet of packing surface area available for vapor-liquid contacting, per cubic foot of tower volume (Lieberman and Lieberman, 2008). The greater the open area and the wetted surface area of packing, the greater the capacity and the separation efficiency of a column, respectively. In addition, other objectives when choosing packing types are to maximize the resistance to mechanical deformation and/or breakage and, especially, to deformation under the weight of the bed, to minimize cost, to maximize fouling resistance, to minimize the sensitivity of a packing in corrosion, erosion, chemical attack, and migration through the support grid openings; to minimize damage during abnormal operation, such as in under weather pressure surges or catch fire at shutdown (Kister, 1992).

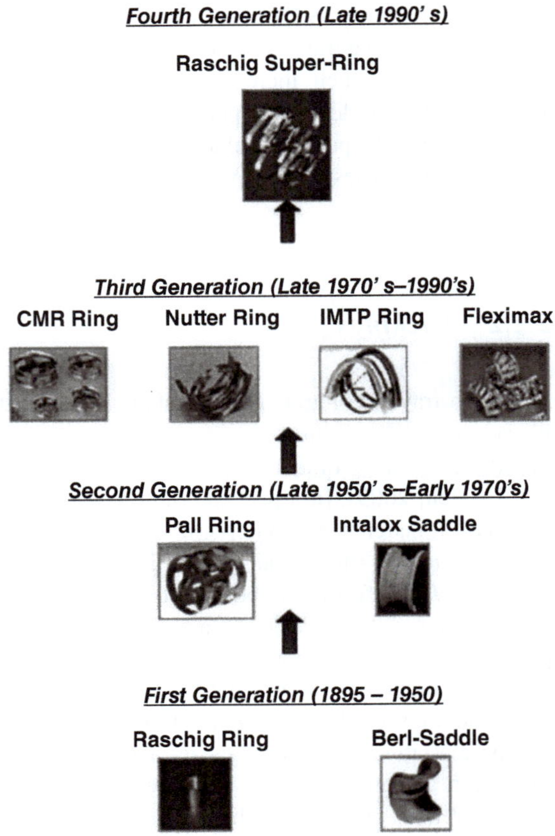

Fig. 2.1 History of the development of characteristic random packings of different generations (Reprinted from Schultes, 2003, with permission from Elsevier B. V.)

2.2.3 Column Internal Performances

To judge the column internals, combination of capacity and efficiency is mainly considered. For a device to be successful, it must have good baseline efficiency and the capacity to meet the process requirements compared to other design alternatives (Hirsch and Pilling, 2010). Capacity and efficiency are interdependent, which causes typical trade-off situations that a column designer needs to deal with when designing a new, or revamping an existing column (Olujić, 2014). They depend on the hydraulic and mass transfer aspects of the column internal operation, such as entrainment, downcomer flooding, and vapor liquid contact. Therefore, these aspects must be understood in the ongoing search for higher performance column internals. For example, in a low-pressure trayed distillation, the vapor densities are low and there will be volumetrically a much larger amount of vapor flowing through the column (Hirsch and Pilling, 2010). In this case, the vapor handling capacity is a critical factor. In these highly vapor loaded systems, the entrainment capacity is typically more important than the downcomer capacity. On the other hand, in high pressure distillation operations, the vapor density is higher and the vapor volume is lower so the liquid handling within the downcomer becomes the most important aspect. When this is well understood, various aspects of the deck and/or downcomer designs and/or others can be modified more easily to achieve the desired performance improvements (Hirsch and Pilling, 2010).

2.2.3.1 Hydraulic

As mentioned above, an understanding of column hydraulic or flooding is needed before any effort of modification. On the tray, the liquid is mixed with vapor to form a foam, which then drains to the tray below through a downcomer after overflowing the weir. Aerated liquid is backed up into the downcomer because of tray pressure drop, liquid height on the tray, and frictional losses in the downcomer apron (Kister, 1992). Flooding can be defined as the inability of the liquid phase to move down the column, which can be the result of an excessive liquid rate, an excessive vapor rate, or a combination of the two (Resetarits, 2014).

In an existing distillation process with a fixed feed composition, there will be a maximum feed flow rate that can be separated to meet certain fixed product specifications (Liu and Jobson, 2004). Nevertheless, an increase in the production capacity can cause flooding in the columns or bottleneck problems. Entrainment (jet) flooding and downcomer backup, which are affected by the internal column loads of the vapor and liquid flows, are two flooding types normally occurring in trayed columns that cause poor tray efficiency, whereas flooding in a packing column occurs when the entire column is filled with liquid. Flooding can also occur locally in a packed bed, hindering the gas phase to move through the bed (Mackowiak and Macckowiak, 2014). The principles of flooding in a column are described in detail in *Distillation Design* (Kister, 1992) and *Working Guide to Process Equipment* (Lieberman and Lieberman, 2008).

2.2.3.2 Efficiency

The overall efficiency, E_o, is defined as the number of equilibrium stages required for separation divided by the actual number of stages required (Wankat, 2012):

$$E_o = \frac{N_{equil}}{N_{actual}} \tag{2.1}$$

Because the efficiency might vary from one stage to stage, the Murphree stage efficiency is commonly used and defined as

$$E_{MV} = \frac{y_n - y_{n-1}}{y_n^* - y_{n-1}} \tag{2.2}$$

where y^* is the composition of vapor in equilibrium with the liquid leaving the stage, and y_n is the actual composition of vapor leaving the stage. The Murphree stage efficiency is the ratio of the change in composition on the stage to the change that would occur on a theoretical stage (Kister, 1992). Distillation stages normally operating between 10% and 90% efficiency have poor performance under a spray regime or when liquid and/or vapor maldistribution occur, or when flooding or dumping (Lieberman and Lieberman, 2008; Kister, 1992).

For the tray type, the efficiency might increase with increasing weir height, length of liquid path, and with a decrease in fractional hole area, hole diameter, and viscosity. In addition, it is also affected by errors in the VLE. At very low relative volatilities ($\alpha < 1.2$), small errors in the VLE have a huge impact on the tray efficiency, for example, at a relative volatility of 1.1, a 3% error in relative volatility gives a 40–50% higher tray efficiency than its true values (Kister, 1992). A loss of tray efficiency can be caused by the high vapor velocity (flooding) or low vapor velocity (turndown). To distinguish between flooding and weeping trays, the tower pressure drop is measured. If the pressure drop per tray, which is expressed in inches of liquid, is more than three times the weir height, then the poor fractionation is due to flooding, whereas the poor fractionation is due to weeping or dumping if the pressure drop per tray is less than the height of the weir (Lieberman and Lieberman, 2008).

In a packed column, the liquid and/or vapor maldistribution is one of the main causes of the failure according to expectations (Kister, 2006). Each tray in a column is inherently a vapor-liquid redistributor, whereas packing does not redistribute liquid or internal reflux (Lieberman and Lieberman, 2008). Therefore, unless the initial reflux distribution is good, the liquid flow distribution through the entire packed bed will be poor. This causes poor fractionation efficiency in a column. Because of this, packed beds should have a limited length as well as liquid collectors and liquid redistributors to redistribute the liquid to the bed below. Unfortunately, conventional liquid collectors, even the chimney type, generate severe initial vapor maldistribution, which is smoothed out within two or three conventional structured packing elements (Olujić, 2007). Therefore, liquid collecting devices should be designed with a minimum pressure drop without compromising the vapor distribution. Furthermore, a gas distributer with shallow beds, large column diameter and low gas velocities is generally required to provide a uniform distribution of the gas at the entrance to the packing (Spiegel and Duss, 2014).

2.2.3.3 Capacity

The capacity here means the volumetric vapor throughput, and the operating capacity is normally set by designers at 0.75–0.90 of the maximum capacity, that is, the point of the onset of flooding (Olujić, 2014). Sieve and valve trays can be considered equal in this respect in the entire range of operating conditions, and bubble cap trays can also handle high vapor loads, but not high liquid loads. The capacity of the trays depends on the tray spacing, liquid load, fractional hole area, and hole diameter (Kister, 1992), whereas the capacity of the various packings depends on the packing size or specific geometric area of the packing. Of course they also depend on the specific vapor and liquid loads, and the physical properties, mainly the densities of vapor and liquid, the surface tension, and to a lesser extent the viscosity of the liquid (Spiegel and Duss, 2014).

2.2.4 *Criteria for Selection of Vapor-Liquid Contactors*

Because the column internals cannot be accessed easily after startup, they are arguably the most important pieces of process equipment. If the column internals malfunction, the distillation column will follow suit and the entire chemical process will suffer. Therefore, the selection of a distillation column internal plays important role and requires an understanding of the purpose of the separation, the required range of vapor and liquid flows, and the physical properties of the vapor and liquid throughput the tower (Pilling and Holden, 2009). These data, together with the vapor-liquid equilibrium and physical models, determine the optimal number of stages and reflux ratios, column diameter and height, flooding percentage, and product purity. The criteria that influence the selection of tray or packing types include the capacity, efficiency, turndown, operating pressure, pressure drop, liquid load, and holdup, sensitivity to fouling and corrosion, construction material, column diameter, cost, as well as know-how/experience (Lockett, 1986; Pilling and Holden, 2009; Olujić et al., 2014).

Pilling and Holden (2009) provided guidelines for when to choose trays and packings for distillation. According to the authors, the choice of trays or packing can be quite complex, so the engineer must consider the impact of many factors in making the best choice for an application. In particular, trays are used in applications with liquid rates of 30 m^3/m^2-h and above, and/or those where solids are present or fouling is a concern, while the greatest advantages of structured packings are usually realized at lower pressures (i.e., <2 atm) and lower liquid rates (i.e., <50 m^3/m^2-h) because of its extremely effective vapor-handling ability. A vacuum to atmospheric pressure column is ideal for structured packing. Random packings are often used in higher-liquid-rate applications, where a lower pressure drop is desired.

Furthermore, at a small column diameter (i.e., <760 mm), trays requiring special mechanical considerations become less practical, whereas packings are almost always used (Pilling and Holden, 2009). Cases where two liquid phases are present in the column provide challenges for both packed and trayed columns. They also claim that in many processes, where thermal degradation is a concern, the use of structured packing can provide two significant advantages. First, structured packing inherently has a very low liquid holdup, so the liquid residence time in the bed is less, which minimizes degradation. Second, the lower pressure drop of structured packing allows the distillation tower bottom sections to be operated at lower pressures and lower temperatures, which also minimizes the degradation potential.

Normally, trays are generally preferred in the distilling of hydrocarbons. On the other hand, trays should not be applied in an offshore plant, such as floating, production, storage, and off-loading facilities (FPSO) and offshore liquefied natural gas (FLNG), which must be able to withstand waves, particularly the distillation and separation columns (Lee et al., 2008). For this reason, columns should be designed to withstand motion, limiting the column height and making packed type column designs suitable (Lee et al., 2012).The walls formed by the corrugated sheets reduce the impact of motion and make structured packing a better choice than random packings (Spiegel and Duss, 2014).

Sieve trays are very good in fouling applications or when solids are present because they are easy to lean. On the other hand, the turndown (i.e., the performance when operating below the designed flow rate) is relatively poor (Wankat, 2012). Valve trays are designed to have better turndown properties than sieve trays. Therefore, they are more flexible when the feed rate varies. Note that tray supports and installation are generally more complicated and expensive than with packing (Resetarits, 2014).

2.3 Retrofit and Debottlenecking by Modifying Column Internal

2.3.1 Using High Capacity/Efficiency Trays

Structured packings are incapable of performing accordingly in separations at higher operating pressures and in the cases where fouling is the main concern. This has led to activities aimed at improving tray type devices to achieve higher capacity at an affordable pressure drop with an acceptable penalty on the separation efficiency (Olujić et al., 2009). Many retrofit projects in distillation have emphasized the replacement of an existing column internal with a high performance column internal, which can improve the separation and maximize capacity. Replacing the existing internals with better ones can be expected to boost the capacity by up to 40%. For new columns designs, high capacity trays can be used to reduce the vessel diameters, heights, or both, while they can increase the capacities, reduce the utilities, and improve the separations for existing columns. These trays can be also used in retrofit projects for improving the column performance providing significant economic rewards, which include an increase in column feed rates for the same energy consumption, a decrease in energy consumption for the same feed rate, an improvement in product purity, a reduction in absorption liquid circulation rates and a corresponding reduction in the regeneration energy cost requirements (Koch-glitsch website).

Note that in the ongoing search for higher performance trays, one first must understand the hydraulic and mass transfer aspects of the tray operation, such as entrainment, downcomer flooding, and vapor liquid contact (Hirsch and Pilling,

2010). When this is well understood, various aspects of the deck and/or downcomer designs can be modified more easily to achieve performance improvements. As described by Bruckert (1968), a conventional cross flow tray has several limitations. The usual discharge weir cannot reasonably be as long as the tray diameter because the downcomer will then occupy half the cross-section area of the tray and will deleteriously encroach on the vapor-liquid contact area and cannot be short, which prevents achieving substantially higher flows with a low, controllable froth height. Another disadvantage of cross-flow trays is the long liquid path to be traversed between the inlet and downcomer. Gravity is the usual driving force for promoting flow along this path, and the resulting hydraulic gradient causes non-uniform vapor ΔP across the tray. The gradient increases at higher liquid rates and the tray becomes unstable and inefficient. A further limitation is the capacity of the downcomer to discharge its clear liquid. Another disadvantage of cross flow trays is their requirement for receiving cups into which the downcomer discharges.

One of main ideals from development of a column internal capacity is increasing the active area, where mass transfer takes place. The UOP Multiple Downcomer (MD) tray (Fig. 2.2) was the first commercially successful tray with an active area under the downcomers (Bruckert, 1968). According to the inventor, this tray provides a distillation system, which is unaffected by the above-

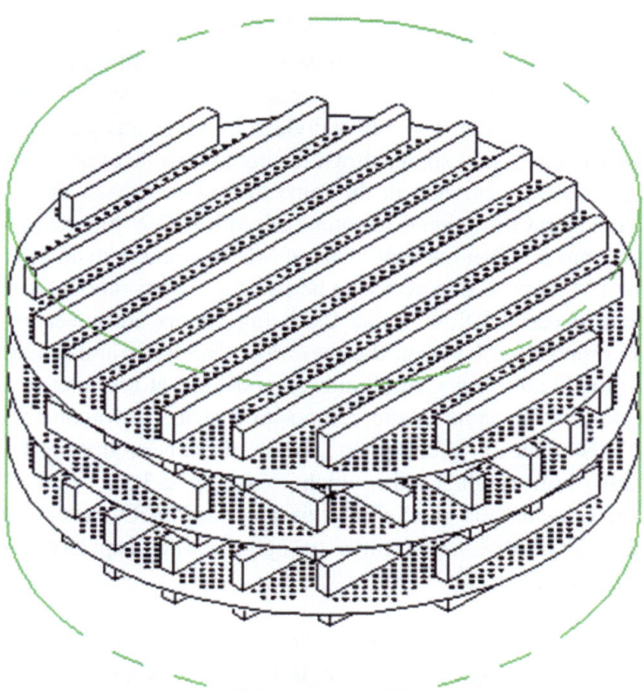

Fig. 2.2 UOP Multiple Downcomer (MD) tray (http://www.distillationgroup.com/questions/question008.htm)

mentioned adverse effects to such an extent that a tremendous increase in liquid flow rate can be accommodated without creating problems of excessive entrainment, downcomer flooding, hydrostatic gradient, or high pressure drop, i.e., to provide more mass transfer capacity per unit column volume with lower power consumption. Another objective is to provide such a system where in the froth height is controllable over wide liquid and vapor flow rates, thereby allowing high rangeability of the system at a reasonable vapor phase pressure drop. Further aims are to provide such a system with an accurately predictable hydraulic behavior and lower tray spacing.

While these MD trays are an advance in the art, a problem exists in that the downcomers give a poor liquid flow distribution on the tray, which results in stagnant areas with regard to liquid flow (Lee and Lemon, 2004). To minimize the inherent problem of MD tray, the UOP developed an Enhanced Capacity Multiple Downcomer (ECMD) (AMT report, 2008; Resetarits and Lockett, 1995). The vapor capacity of MD trays is increased by the provision of vapor directing slots in the tray deck with the slots being oriented toward the closest downcomer to reduce the froth height. The slots located in the relatively narrow deck area between downcomers are divided into two sets that point in opposite directions. In addition, baffles are provided above the downcomer inlets, which preferably extend upward to slightly below the lower surface of the next higher tray, with the upper edge of the baffles being notched to accommodate the downcomers of this next higher tray.

Furthermore, the Parallel Flow Multiple Downcomer (PFMD) tray is the latest generation of UOP trays to employ parallel flow to improve the efficiency (Brandner and Andre, 2012). The tray is divided into four quadrants, each separated by a downcomer. The liquid spirals down the column so that the liquid in each column quadrant is always flowing in the same direction on each tray. According to the authors, the spiral flow liquid is combined with MD™ tray technology so that the PFMD tray has a higher capacity than an MD tray and with an efficiency that is greater than an MD tray and expected to be equal to a conventional tray.

The benefit of parallel flow is that the rising vapors are never exposed to the same liquid composition between the two trays (Brandner and Andre, 2012). On conventional distillation trays, the liquid composition at the inlet to the downcomer on one tray is the same as the composition at the downcomer outlet on the tray below. This creates a "pinch point", where there is no driving force for mass transfer. Parallel flow eliminates this pinch point (Fig. 2.3).

In addition to the unique parallel flow design, the PFMD trays combine the features of other UOP distillation trays to optimize the capacity and efficiency (Brandner and Andre, 2012):

- Similar toother trays in the MD family, PFMD trays have long weir lengths and no seal pans for high capacity and low pressure drop.
- Directional slots from UOP slotted sieve trays are used to form a preferential liquid flow direction, eliminate the stagnant areas on the tray deck, and reduce the froth height for a higher vapor handling capacity and separation efficiency.

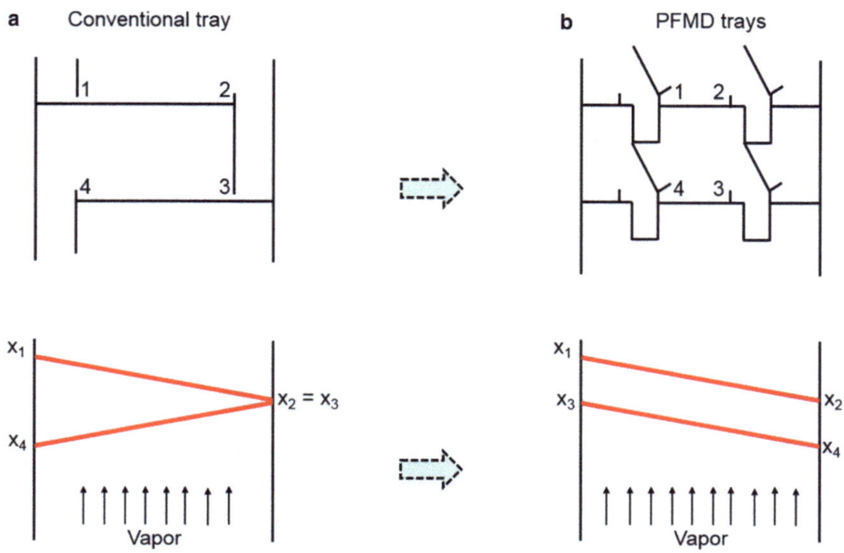

Fig. 2.3 Parallel flow multiple downcomer trays

- Bubble promoters are used at the inlet to each tray deck to initiate bubbling, reduce the liquid head, and prevent weeping at the inlet. This results in a lower pressure drop, and higher efficiency and capacity.
- A sloped downcomer design allows for a deeper downcomer for higher capacity and improved operating range.

The combination of high capacity, high efficiency, and low pressure drop make PFMD trays a great solution for existing bottlenecked columns in expansion revamps without a loss of separation efficiency and product purity, and for installed cost savings in lower-pressure grassroots columns (Brandner and Andre, 2012).

AMT International, INC. developed a radically different version of a multi-downcomer tray, called the ADV® LD (Lattice Downcomer) tray in early 2001 (Fig. 2.4) (AMT report, 2008). The key claim of the ADV® LD development is that multiple (rectangular-shaped) downcomer boxes are arranged in such a way that all downcomers are connected and equalized in a "gridlike" assembly, as shown in Fig. 2.4c. The advantages of ADV® LD include uniform liquid flow distribution (resulting from perfect radial flow pattern), good liquid mixing from tray to tray, and a longer liquid flow path than the MD trays of the prior art (Lee and Lemon, 2004).

A dual-flow tray is a sieve tray with no downcomers (Kister, 1992). In this tray, vapor and liquid flow counter-currently through the perforations with liquid continuously weeping through the holes, hence its low efficiency. The tray froth height decreases rapidly when vapor velocity is reduced, causing a further decrease in efficiency upon turndown. Owing to the absence of downcomers,

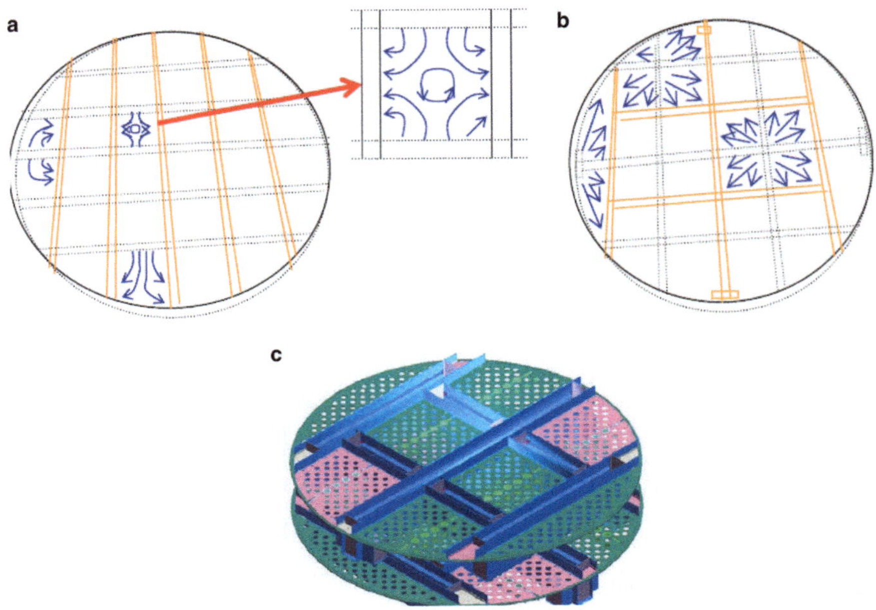

Fig. 2.4 (a) Multiple downcomer tray; (b) ADV® LD tray; and (c) ADV® LD tray (www.amtintl.com/adv-ld-tray.pdf)

dual-flow trays provide a higher tray area, and have a greater capacity than any of the common tray types. This makes them an ideal revamp tool if some efficiency can be sacrificed. The absence of downcomers, and the larger open areas, render dual-flow trays the most suitable for handling highly fouling services, slurries, and corrosive services. Dual-flow trays are also the least expensive to make, and easiest to install and maintain.

In addition, a gas-liquid contacting apparatus was proposed, in which the area under the downcomer was converted to the active area (Fig. 2.5) (Chuang and Everatt, 1985). This apparatus produces a frothing layer on the trays, in which the perforated, inclined, liquid distributing, gas entry plates are immersed, and which gives a good liquid flow distribution across the tray so that a high efficiency of contact between the gas and the liquid can be achieved.

On the other hand, this is still a problem area from the point of view of liquid passing directly down through the perforations intended for the upward passage of vapor. Therefore to some extent, the desire to attain the greatest contact efficiency, by a perforation as large as a part of the deck surface as possible is at odds with the desire to avoid weeping (Sauter et al., 1997).

An ADV™ pinnacle performance tray shown in Fig. 2.6 was developed with many advantages over conventional trays (AMT pacific website), such as up to 40% greater throughput capacity, up to 10% better fractionation efficiency, up to 20% lower pressure drop, greater turn down ratio, minimum modifications for

Fig. 2.5 Hanging downcomer tray.

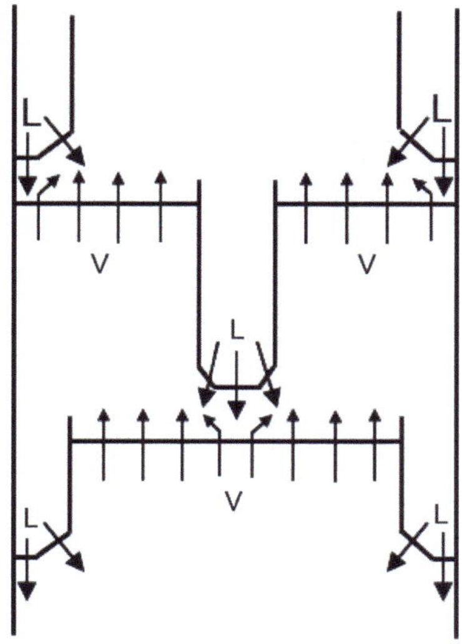

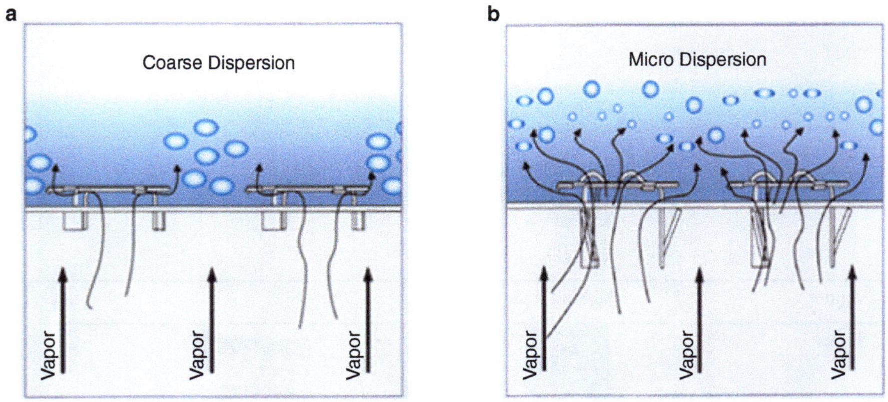

Fig. 2.6 Operation mechanism of (a) conventional valve tray and (b) ADV™ tray (http://www.amtpacific.com/eng/)

debottlenecking, easier field installation of trays, quick payback of debottlenecking costs, and lower initial investment for new towers, which can be applied into oil refineries, petrochemical and chemical plants.

Recently, Kochglitsch developed high capacity/efficiency trays, such as TRITON®, SUPERFRAC, and ULTRA-FRAC® trays (Table 2.1). Currently, all vendors market many types of high performance trays/packings. These vendors also support useful rating tools, such as the KG-Tower™ (Koch-glitsch) or Sulcol™ (Sulzer), which can be used initially to check the performance of high capacity trays/packings. Table 2.1 lists several typical high performance trays.

Table 2.1 Several typical high performance trays/packings

Vendor	Tray trade name			
Koch -Glitsch	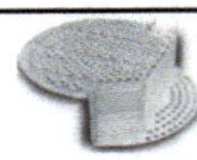 TRITON®	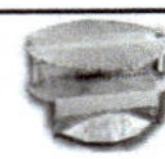 SUPERFRAC	 ULTRA-FRAC®	 Minivalve trays
	 PROVALVE™			
Sulzer	 MVG/SVG trays	 VGPlus™ trays	 Hi Fi Plus trays	 Con Sep trays
Raschig GmbH	 CoFlow trays			
AMT	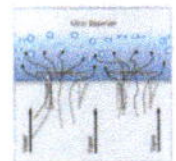 ADV™ trays			
UOP	 PFMD			

Vendor	Packing trade name		
Sulzer	 MellapakPlus	 BXPlus	 Mellagrid
Koch -Glitsch	 Flexipac HC	 AX	
Montz	 Montz-pak		
Raschig GmbH	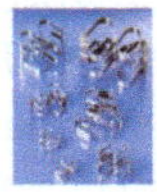 Raschig Super-Ring		

http://www.koch-glitsch.com/masstransfer/pages/high_performance_trays.aspx

2.3.2 *Using High Capacity/Efficiency Packings*

The performance of trays, random and structured packings in terms of the capacity and efficiency depend directly on the pressure drop. In particular, the upward flow of the vapor phase is associated with a certain pressure drop across the column internals, that of trays being much higher than that exhibited under the same conditions by random and particularly structured packings (Olujić, 2014). A larger pressure drop produces a larger pressure and, consequently, a larger temperature at the bottom of the column. If excessive, it can deleteriously affect the separation (reduced relative volatility, bottom product degradation by decomposition or polymerization, and/or need for a hotter heating medium). Therefore, in many cases, packings are favored. Together with greater sustainable development, they are also under efforts to improve the performance.

In the case of packings, the capacity is related directly to the packing size, i.e. specific geometric area (Olujić, 2014). The larger the random packing size or corrugation dimensions (smaller specific geometric area), the larger the capacity. Note that the efficiency in that case is normally reduced. The interdependency of the capacity and efficiency causes a trade-off situation that design engineers should be aware of.

A large number of new random packings with a relatively large open area in the ring wall considered as the third generation were introduced on the market, including NOR-Pac, Hiflow rings, Top-Pak, ENVIPAC, Dtnpac, VSP ring Ralu-Flow, Tri-Pak, ceramicR-Pac and SR-Pac, CMS-cascade mini rings, and Intalox metal packing IMTP (Mackowiak and Macckowiak, 2014). They are called lattice-type packings because the open structure has a very wide range of applications. Further development in this area of lattice random packings, four generation, such as Raschig Super Ring, was introduced to provide an exceptionally low pressure drop, high capacities, and excellent mass transfer efficiency (Schultes, 2003).

While considering the hydraulic behavior of the conventional structured packing with different corrugation angles, the overall pressure drop was found to consist of three major components: gas-liquid interaction at the interface along the flow channels, gas–gas interactions at the crossings of the oppositely-oriented flow channels of the adjacent corrugated sheets, and the pressure loss associated with an abrupt change in the flow direction at the transitions of the packing elements (Olujić, 2007). The latter two, which are responsible for, say 80% of the total pressure drop in packings with a corrugation angle of 45°, do not contribute greatly to mass transfer (Olujić et al., 2003). This means that reducing these pressure drop components appropriately can lead to an increase in capacity without adversely affecting the mass transfer efficiency or vice versa. By modifying the corrugation angle from 45° to 60° to limit the pressure drop, the capacity increased considerably; however, this was at the expense of roughly equivalent loss of efficiency (Olujić, 2007).Therefore, new challenge for packing manufacturers was to arrive at a structure that will combine the efficiency of a 45° packing with the capacity of a 60° packing.

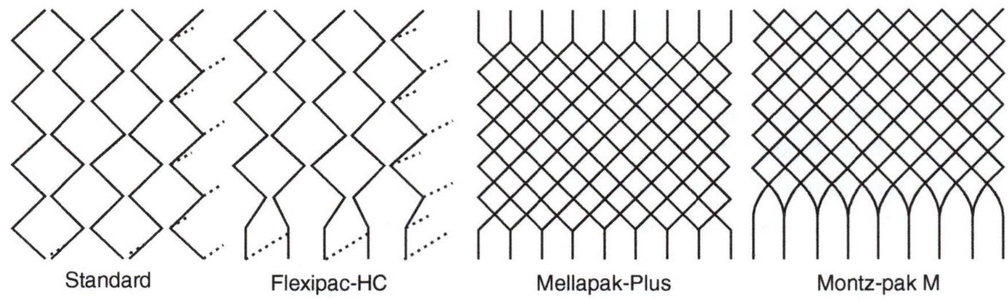

Fig. 2.7 Schematic diagram of the proprietary designs of the new generation of high-performance structured packings (Reprinted with permission from Olujić et al., 2001. Copyright (2001) American Chemical Society)

Replacing the trays or standard packings with high performance packing allows greater capacity and separation that prevents the damaging the product. Furthermore, using the packing type, a generally shorter tower height, makes it possible to accommodate more equilibrium stages than the original trayed column. This leads to a substantial decrease in reflux ratio and energy requirement, i.e., the column is possible to increase the capacity after retrofit. Many existing columns have been retrofitted using a new generation of high-performance random packings, such as Raschig Super-Ring (Raschig GmbH), or high performance sheet packings, such as MellapakPlus (Sulzer), Flexipac HC (Koch-glitsch), and Montz-Pak (Montz) with different corrugation angles at the ends of each layer, as shown in Fig. 2.7, or high-performance gauze packings, such as BXPlus (Sulzer), AX (Koch-glitsch), or high-performance grid packings such as Mellagrid (Sulzer), which allow an up to 30% increase in column capacity (Spiegel and Duss, 2014; Olujić et al., 2001; Schultes, 2003; Bender and Moll, 2003; Faessler et al., 2007) (Table 2.1).

2.3.3 Column Internal Structure Modifications

Many times, the column is operated near its flood point, and engineers are faced with the requirement of a 5–20% capacity increase in their fractionation trains. In addition to replacing the existing internals with high-capacity and/or high-efficiency internals, improving the existing internals or structure can be considered. For a high liquid rate, the one-pass tray is changed to multi pass cross flow trays (Fig. 2.8) (Coker, 2010). Change cross flow trays to parallel-flow trays, which can increase the efficiency by up to 20% (Fig. 2.9) (Resetarits, 2014).

The use of segmental downcomers at high liquid loads, which extend the weir length, reduces the pressure drop and downcomer backup, improves the liquid distribution on the tray and increases the tray capacity (Karpe, 2009). They are used in revamps for increasing the downcomer width. Furthermore, the sloped

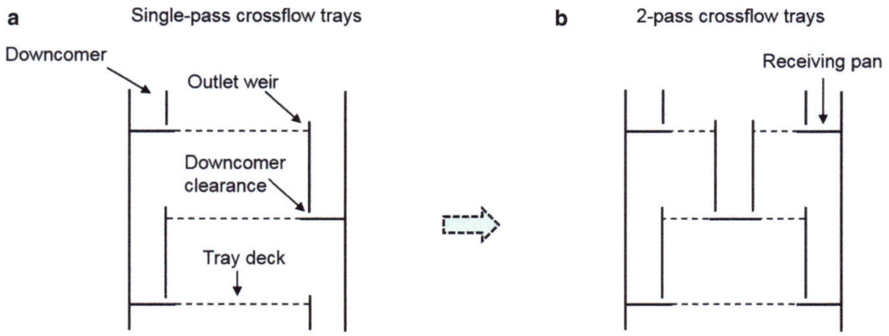

Fig. 2.8 Changing the single-pass crossflow trays to 2-pass parallel-flow trays

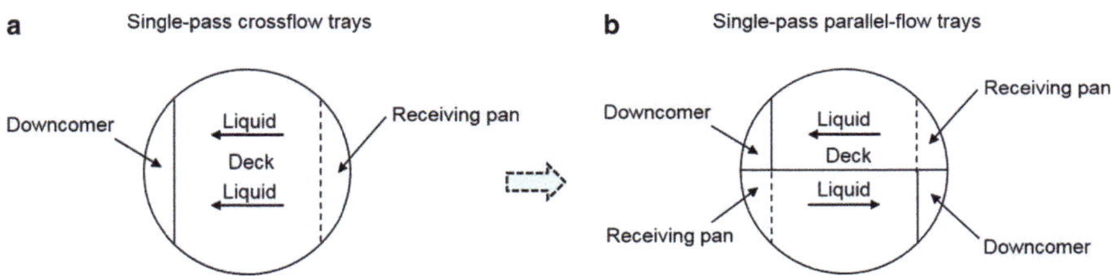

Fig. 2.9 Replacing the single-pass crossflow trays by single-pass parallel-flow trays

Fig. 2.10 Sloped downcomer

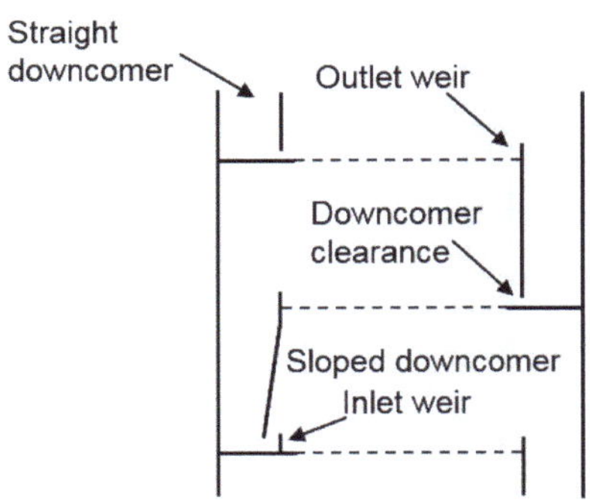

downcomer is used to slightly increase the active area and improve the downcomer separation without wasting the active area on the tray below to increase the tray capacity (Fig. 2.10).

Besides, radial tipped downcomers are used for high liquid loads (Fig. 2.11). The smooth radial tip of the downcomer allows a lower pressure drop through the downcomer clearance and reduces downcomer backup. This leads to higher capacity (Karpe, 2009).

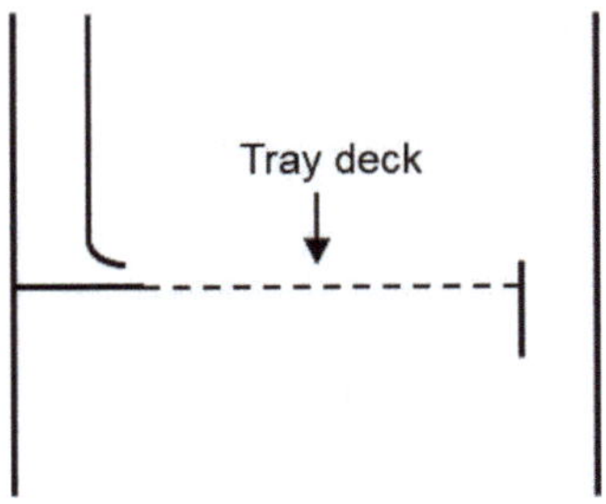

Fig. 2.11 Radial tipped downcomer

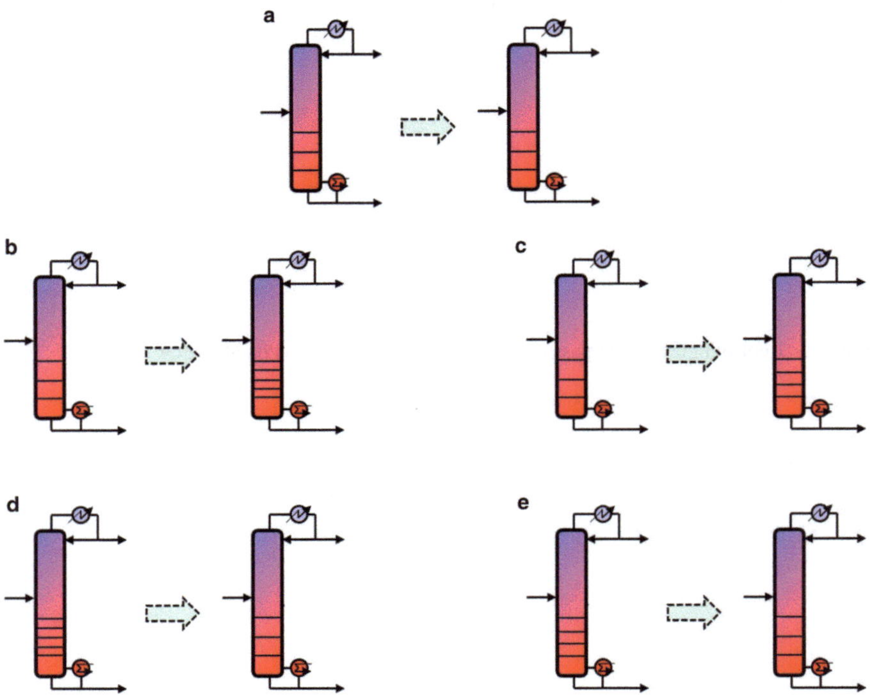

Fig. 2.12 Trayed column retrofit with (a) 1 for 1, (b) 2 for 1, (c) 3 for 2, (d) 1 for 2, and (e) 2 for 3 replacement

The jet capacity increases with increasing bubbling and free area, or tray spacing increases, or open area increases, or hole size decreases. Figure 2.12a shows the simple one for one replacement; as an example in below, 92 high-capacity trays can be placed in the existing column to replace 92 original trays. Furthermore, adding more trays, which reduces the tray spacing, might decrease the reflux ratio, energy requirement, vapor and liquid traffic, and flooding % (shown in Fig. 2.12b and 2.12c). On the other hand, the closer the tray spacing, the less the tray flexibility (Coker, 2010). Retrofits sometimes fail, particularly when the design engineer learns too late that the transition piping and ancillary equipment (e.g. heat exchangers, pumps, receivers, etc.) also requires revamping (Resetarits, 2014).

Industrial columns have used spacing's from 150 to 1200 mm (Resetarits, 2014). A low tray spacing also enhances the tendency of entrainment and downcomer backup flooding. In other words, the jet capacity increases with increasing tray spacing (Chen, 2004). Therefore, increasing the tray spacing might increase the capacity (shown in Fig. 2.12d and 2.12e); however, it increases the column height or reduces the separation efficiency. This method is effective when the existing column has a large number of trays, which can remove several flooded trays to reduce the flooding value in that area, while the purity and duty show almost no change. In some rare cases, only a few trays are flooded when increasing the capacity; one for two or two for three replacements can considered.

Enlarging the hole area can reduce the hole vapor velocity, dry tray pressure drop and entrainment, which can improve the capacity slightly (Kooijman and Taylor, 2014; Yanagl and Sakata; 1982). However, this can cause a slight decrease in tray efficiency.

In a packed tower, the column capacity can be increased using larger packing and increasing the open area (or a bigger crimp size for structured packing) (Lieberman and Lieberman, 2008). On the other hand, the reduction in the wetted area of the packing will reduce the separation efficiency. Installing a properly designed, efficient liquid distributor will definitely enhance the separation efficiency, but at the expense of the vertical height of the tower, which could otherwise be devoted to a greater packing height.

2.3.4 Industrial Cases

Numerous industrial cases, in which the existing column internal was modified or replaced with high performance trays/packings, have been reported. They are employed to improve the performance from distillation columns to absorbers and strippers (Schultes, 2003), from vacuum conditions (Shilkin, 2010) to high-pressure conditions (Summers et al., 1995). For example, in 1992, Chevron Chemical Company placed 325 of UOP's Enhanced Capacity Multiple Downcomer (ECMD) trays, which represents a combination of the Multiple Downcomer (MD) tray and Slotted Sieve technologies, in its large propylene/propane splitter at Port Arthur, Texas, resulting in a 40% increase in capacity to approximately 124,000 lb/hr (Summers et al., 1995). Another example is that UOP sold a set of Parallel Flow Multiple Downcomer (PFMD) trays in 2011 for use in an Isomar Unit Deheptanizer for the Arabian Fibers Industrial Company (IBN RUSHD) IR-2 aromatics debottlenecking project in Yanbu, Saudi Arabia (Brandner and Andre, 2012). The objective of the revamp was to increase the paraxylene product capacity and maximize benzene production. This marks the first commercial sale of PFMD trays. Many retrofit projects, where the packing types were introduced, such as a Raschig Super Ring, was were successfully in many retrofitted columns such as Toluene-di-isocyane process (BASF), vacuum distillation columns (Degussa AG), C_3 splitter (Targor Lillebonne), which led to higher capacity, lower operating costs with a small tendency to plugging and low foaming (Schultes, 2003).

Although the use of these trays/packings can improve the efficiency and/or capacity, engineers should be aware that installing these new internals can be costly and may require substantial changes. Therefore, it is not always necessary to switch to better internals if the existing internals can be improved significantly (Williams, 1998). Furthermore, many plants have had successes, whereas others have had repeated failures when replacing the existing column internal with high-capacity trays. Often, the difference between success and failure is understanding the performance limits and compromises in advance (Stoley, 2001). Effective application of high-capacity trays requires more precise knowledge of the operating conditions and flexibility requirements, i.e., greater attention to design detail and careful installation.

2.4 Case Study

2.4.1 Existing Conventional Distillation

A deisobutanizer, possessing 92 theoretical trays (Amminudin and Smith, 2001; Long and Lee, 2013), was designed and operated at 6.5 bar (Fig. 2.13) because commercial isobutane (iC_4) can be condensed with cooling water under this

Fig. 2.13 Simplified flow sheet illustrating the existing deisobutanizer column

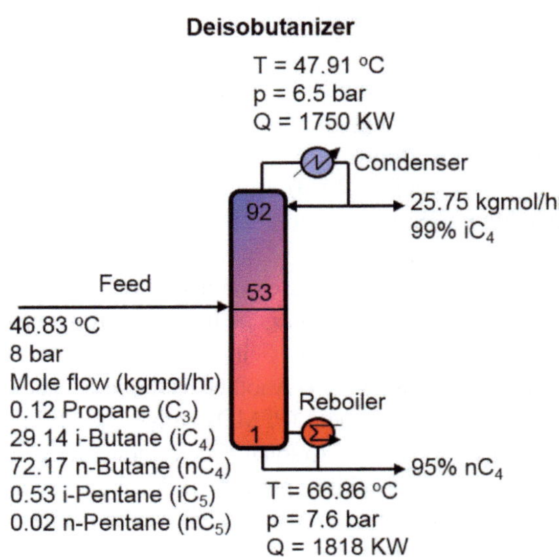

pressure (Klemola and Ilme, 1996). The simulations were performed using the commercial simulator, Aspen Hysys V8.4. The Peng-Robinson equation of state was used to predict the vapor-liquid equilibria of these simulations (Aspen Technology, 2009). The flooding percentage of the modeled columns was determined by both Aspen Hysys V8.4 and KG-Tower^{TM} V5.2 software using a rating mode simulated using the internal specifications of the columns, such as the type of trays, column diameter, tray spacing, number of passes, hole diameter, open area, weir height, downcomer clearance, and the downcomer top and bottom widths. Note that in Aspen Hysys, the bottom downcomer widths can only be specified when sloped downcomers are selected from the downcomer type drop-down list. In both software packages, the weir length is calculated automatically after the downcomer width is specified. Table 2.2 lists the column specifications and hydraulic for the existing deisobutanizer column. Note that the weir length is typically 60–75% of the tray diameter (Wankat, 2012; Coker, 2007), while in this case, that is about 78.6% to avoid downcomer flooding. The base case simulation model showed that the energy consumption of the deisobutanizer was 1818 KW.

Figure 2.14 presents the flooding, vapor and liquid profiles of existing deisobutanizer. In this column, the vapor and liquid traffic are high in the bottom section, which results in a higher flooding profile. On the other hand, the maximum vapor and liquid flow might not indicate adequately which stages of the column have the maximum flooding %. In particular, the vapor profile gains the greatest value at the 17th tray, whereas the highest flooding % (86.9%) is observed at the 52nd and 53rd trays.

After the column is converged, the information of the vapor and liquid net flow rate, as well as the physical properties for each tray are available. They are input to the KG-Tower^{TM} to determine the flooding percentage. As shown in Table 2.2, the maximum flooding calculated by KG-Tower^{TM} is 85.6%, which is in good agreement with the result simulated by Aspen Hysys.

Table 2.2 Column specifications and hydraulics of the existing column

Specification	Existing deisobutanizer	
Number of trays	92	
Tray type	Sieve	
Column diameter (mm)	1100	
Number of flow paths	1	
Tray spacing (mm)	457	
Hole diameter (mm)	6.35	
Open area (%)	10.08	
Downcomer top and bottom widths (mm)	210.0	
Weir height (mm)	864.6	
Downcomer clearance (mm)	38.1	
	Aspen Hysys	**KG-Tower^{TM}**
Maximum flooding (%)	86.9	85.6

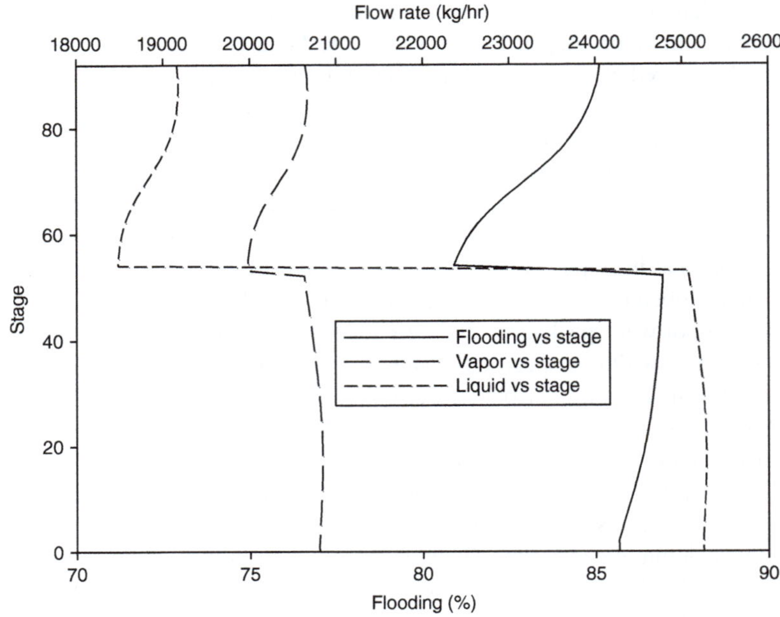

Fig. 2.14 Flooding, vapor and liquid flow profiles

Table 2.3 Flooding values

	Tray type			
	Sieve	TRITON®	SUPERFRAC®	ULTRA_FRAC®
Maximum flooding (%)	85.6	65–68	65–68	40–52
Total tray DP (mm Hg)	3.16	2.29	2.29	3.72

2.4.2 *Retrofitted Distillation*

In this case, a simple one for one replacement is employed, i.e., 92 high-capacity trays can be placed in the existing column to replace 92 original trays. As shown in Table 2.3, the capacity of the existing deisobutanizer can be increased because the flooding percentage is decreased from 85.6 to less than 68% by replacing the existing sieve tray by the high performance trays with a simple one for one replacement.

2.5 Concluding Remarks

Nowadays, all vendors market many types of high performance trays/packings aggressively. These vendors also support useful rating tools, such as KG-Tower™ (Koch-glitsch) or Sulcol™ (Sulzer), which can be used initially to check the performance of high capacity trays/packings. Many retrofit projects in distillation have emphasized modifying or replacing existing column internals with high performance column internals, which can promote separation as well as maximize the capacity and reduce energy efficiency. Note that while many retrofit projects were successful, others have had repeated failures when replacing the existing column internal with a high-capacity one. The development is still continued and it is expected to achieve a more innovative column internal, which can improve the capacity, efficiency and energy efficiency of distillation.

References

Amminudin KA, Smith R (2001) Design and optimization of fully thermally coupled distillation columns. Part 2: Application of dividing wall columns in retrofit. Trans. IChemE 79:716–724

AMT report (2008) AMT Lattice downcomer tray. www.amtintl.com/adv-ld-tray.pdf

Aspen Technology (2009) Aspen HYSYS thermodynamics COM interface. Version number V7.1

Bender P, Moll A (2003) Modifications to structured packings to increase their capacity. Chem Eng Res Des 81:58–67

Brandner K, Andre R UOP website, UOP Sells First PFMD™ Tray in aromatics revamp. http://www.uop.com/uop-sells-pfmdtm-tray-aromatics-revamp/

Bruckert W (1968) Vapor liquid contacting system and method. U.S. Patent 3,410,540, 12 November

Chen GX (2004) Optimum column design for capacity and efficiency. AIChE Spring National Meeting New Orleans, Monday, April 26

Chuang KT, Everatt AE (1985) Gas liquid contacting apparatus. U.S. Patent 4,504,426, 12 March

Coker AK (2007) Ludwig's applied process design for chemical and petrochemical plants, vol 1, 4th edn. Elsevier, Oxford

Coker AK (2010) Ludwig's applied process design for chemical and petrochemical plants, vol 2, 4th edn. Elsevier, Oxford

Darakchiev S, Semkov K (2008) A study on modern high-effective random packings for ethanol-water rectification. Chem Eng Technol. 31:1039–1045

Faessler P, Kolmetz K, Seang KW et al (2007) Advanced fractionation technology for the oleochemical industry. Asia-Pac J Chem Eng 2:315–321

Górakand A, Olujić Ž (2014) Distillation: equipment and processes, 1st edn. Elsevier, New York

Hirsch S, Pilling M (2010) Improved high performance trays. In: de Haan AB, Kooijman H, Górak A (eds) Distillation Absorption 2010. Eindhoven University of Technology, Eindhoven, pp 629–634

Jödecke M, Friese T, Schuch G et al (2006) The sandwich packing—a new type of structured packing to increase capacity and mass transfer of distillation Columns. IChemE Symp Series 152:786–789

Jödecke M, Friese T, Schuch G (2007) Hydraulic investigations to increase the capacity of trays in distillation columns. In: Kister H, Pritchard M (eds) Proc. topical distillation conf. AIChE, Washington, p 391–397

Karpe P (2009) Advanced distillation: operation, troubleshooting and energy reduction, November, Kuala Lumpur

Kashani N, Siegert M, Sirch T (2005) A new kind of column packing for conventional and reactive distillation – the sandwich packing. Chem Eng Technol. 28:549–552

Kister HZ (1992) Distillation design. McGraw-Hill, New York

Kister HZ (2006) Distillation troubleshooting. Wiley, Hoboken
Klemola KT, Ilme JK (1996) Distillation efficiencies of an industrial-scale i-butane/n-butane fractionator. Ind Eng Chem Res 35:4579–4586
Kooijman HA, Taylor R (2014) Distillation of bulk chemicals. In: Górak A, Schoenmakers H (eds) Distillation operation and applications, 1st edn. Elsevier, New York, p 191–253
Lee YM, Cho TI, Lee JH, et al (2008) Trends and technologies in LNG carriers and offshore LNG facilities. Offshore Technology Conference, Houston, 5–8 May
Lee AT, Lemon FW (2004) Gas liquid contacting apparatus. U.S. Patent 6,746,003, 8 June
Lee SG, Long NVD, Lee MY (2012) Design and optimization of natural gas liquefaction and recovery processes for offshore floating liquid natural gas plants. Ind Eng Chem Res. 51:10021–10030
Lieberman NP, Lieberman ET (2008) A working guide to process equipment, 3rd edn. McGraw-Hill, New York
Liu ZY, Jobson M (2004) Retrofit design for increasing the processing capacity of distillation columns: 1. A hydraulic performance indicator. Trans IChemE 82:3–9
Lockett MJ (1986) Distillation tray fundamentals, Cambridge University Press, New York
Long NVD, Lee MY (2013) A novel NGL (natural gas liquid) recovery process based on self-heat recuperation. Energy 57:663–670
Mackowiak J, Macckowiak JF (2014) Random packings. In: Górakand A, Olujić Ž (eds) Distillation: equipment and processes, 1st edn. Elsevier, New York, p 85–144
Olujić, Ž (2007) Packed columns: high capacity internals, Elsevier
Olujić Ž (2014) Type of distillation column internals. In: Górakand A, Olujić Ž (eds) Distillation: equipment and processes, 1st edn. Elsevier, New York, p 1–34
Olujić Ž, Jansen H, Kaibel B et al (2001) Stretching the capacity of structured packings. Ind Eng Chem Res. 40:6172–6180
Olujić Ž, Jödecke M, Shilkin A et al (2009) Equipment improvement trends in distillation. Chem Eng Process. 48:1089–1104
Olujić Ž, Kaibel B, Jansen H et al (2003) Distillation column internals/configurations for process intensification. Chem Biochem Eng Q. 17:301–309
Penciak J, Niewoudt I, Spencer G (2006) High-performance trays: getting the best capacity and efficiency. IChemE Symp Series 152:311–316
Pilling M, Holden BS (2009) Choosing trays and packings for distillation. Chem Eng Prog 105 (9):44–50
Resetarits M (2014) Distillation trays. In: Górakand A, Olujić Ž (eds) Distillation: equipment and processes, 1st edn. Elsevier, New York, p 35–84
Resetarits MR, Lockett MJ (1995) Multiple-downcomer fractionation tray with vapor directing slots and extended downcomer baffles. U.S. Patent 5,382,390, 17 Jan
Sauter JR, Hauser RP, Harris J (1997) Fractionation trays. U.S. Patent 5,618,473, 8 April
Schultes M (2003) Raschig Super-Ring a new fourth generation packing offers new advantages. Chem Eng Res Des 81:48–57
Shilkin A, Heinen K, Großmann C et al (2010) On the development of an energy efficient packing for vacuum distillation. Distillation absorption, 653–658
Smith R (2005) Chemical process design and integration. Wiley, West Sussex
Spiegel L, Duss M (2014) Structured packings. In: Górakand A, Olujić Ž (eds) Distillation: equipment and processes, 1st edn. Elsevier, New York, p 35–84
Stoley A (2001) High-capacity trays, http://www.distillationgroup.com/questions/question008.htm
Strigle RF Jr (1994) Packed tower design and applications: random and structured packings, 2nd edn. Gulf Publishing Co, Houston
Summers DR, McGuire J, Resetarits R et al (1995) High-capacity trays debottleneck Texas C3 splitter. Oil & Gas Journal, 93. http://www.ogj.com/articles/print/volume-93/issue-45/in-this-issue/refining/high-capacity-trays-debottleneck-texas-c3-splitter.html.

Wankat PC (2012) Separation process engineering, 3rd edn. Prentice Hall, Upper Saddle River
Weiland R, DeGarmo J, Nieuwoudt I (2005) Converting a commercial distillation column into a research tower. Chem Eng Prog. 101:41–46
Williams JA (1998) Optimize distillation system revamps. Chem Eng Prog 94(3):23
Yanagl T, Sakata M (1982) Performance of a commercial scale 14% hole area sieve tray. Ind Eng Chem Process Des Dev. 21:712–717

Chapter 3
Promising Retrofit Technologies for Single Column

3.1 Introduction

To improve the energy efficiency, the heat integration concept was introduced, whose basic idea is to utilize the heat from the hot process streams to heat the cold process streams and minimize the heating and cooling requirements. The integration is important because it can conserve energy. To obtain a physically realizable estimate of the required heating and cooling duties, a possible temperature driving force must exist between the hot and cold streams, that is, any heat exchanger must satisfy the second law as well as the first law (Douglas, 1988). The flow rate, composition as well as temperature of each stream need to be known before considering heat integration. To satisfy the second law, compression work and/or raising of column pressure can be applied for upgrading the low-level energy. Furthermore, distillation intensification techniques, such as reactive distillation, cyclic distillation and innovative hybrid configurations can be considered to improve the performance of single column. In this chapter, only heat integration configurations and cyclic distillation are mentioned, whereas other configurations are analyzed and evaluated in Chap. 5.

3.2 Heat Pump Assisted Distillation

3.2.1 Introduction

Distillation is a well-defined separation process consuming the largest portion of energy in industry, making this process a major concern for sustainable development in industrially developed countries (Dejanović et al., 2014). This physical process separates liquid mixtures based on the differences in boiling point or relative volatility of the constituent components (Kiss, 2014). The highest heat is supplied to the reboiler to generate vapor flow, which is discarded mostly in the overhead condenser

N.V.D. Long, M. Lee, *Advances in Distillation Retrofit*, DOI 10.1007/978-981-10-5901-8_3

(Matsuda et al., 2011). Heat pumps in distillation allow the heat of condensation released at the condenser to be used for evaporation in the reboiler (Bruinsma and Spoelstra, 2010). This is an economical way of conserving energy when the difference in temperature between the overhead and the bottom of the column is small and the heat load is high (Kiss et al., 2012; Long and Lee, 2014).

Several heat pump concepts have been proposed to retrofit for improving the energy efficiency of distillation columns. The vapor compression (VC) heat pump (Fig. 3.1a) uses a working fluid, which is evaporated at the condenser, compressed to a higher (saturation) temperature, condensed in the reboiler, and cooled by expansion over a throttle valve to a (saturation) temperature below the condenser temperature (Bruinsma and Spoelstra, 2010). The selection of a proper working fluid is an important degree of freedom in the design. An industrial example of a VC is ethylene–ethane separation using propylene as the working fluid (Zimmerman and Walzl, 2009). In some cases, the distillate vapor cannot be compressed, the VC is an attractive option in some novel heat pump systems. In addition, VC is particularly beneficial when dealing with corrosive and fouling compounds (Göktun, 1995).

The mechanical vapor recompression (MVR) heat pump (Fig. 3.1b) is applied widely in the separation of mixtures with similar boiling points. The overhead vapor is pressurized using a compressor to such a level that it can be condensed at a higher temperature supplying sufficient heat to the reboiler. This leads to a major decrease in the energy requirement of reboiler and condenser (Moser and Schnitzer, 1985; Ranade and Chao, 1990; Mizsey and Fonzo, 1992; Annakou and Mizxey, 1995). The use of mechanical compression as a heat pump is economical, mainly in separating low relative volatility mixtures, where small compression ratios and consequently small compressor duties are needed due to the small temperature difference between the top and bottom of the column (Long and Lee, 2014). The advantage over a VC is that the condenser in a MVR is smaller and that the temperature lift is approximately 10 K lower because heat is exchanged only once. This results in higher thermodynamic efficiency (Bruinsma and Spoelstra, 2010).

In a bottom flashing heat pump (Fig. 3.1c), the bottom stream from the column is divided into two streams: the final product stream and another that is expanded by a valve to decrease its temperature and allow heat exchange with the top stream

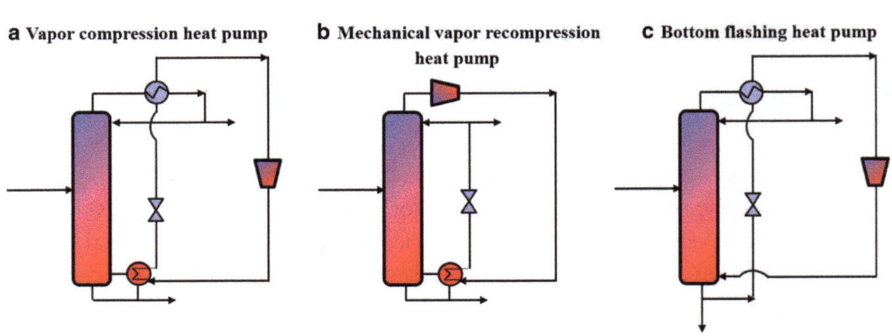

Fig. 3.1 Schematic diagram of the heat pump systems: (a) vapor compression; (b) mechanical vapor recompression; and (c) bottom flashing

before being recompressed to the column pressure (Fonyo et al., 1995; Long and Lee, 2013a). Fonyo et al. (1995) used this system to separate iC_4 and normal butane (nC_4), whereas Long and Lee (2013a) combined this technique with a bottom dividing wall column to integrate the debutanizer and deisobutanizer, and reported remarkable energy savings.

In addition to these three main kinds of heat pumps, several heat pump concepts have been proposed to upgrade the low-level energy and re-use it in distillation, such as thermal vapor recompression (TVR), compression–resorption heat pumps (CRHPs), absorption heat pumps (AHPs), thermo-acoustic heat pump (TAHP), Stirling cycle, Vuilleumier cycle, Brayton cycle, Malone cycle, solid-sorption cycle, and internally heat integrated distillation column (HIDiC), which are described in detail in Kiss's book (Kiss, 2013). This book also presents a novel selection scheme based on an extensive literature survey taking into consideration the most promising energy efficient distillation technologies that currently feature shorter implementation times. In this chapter, applications of a heat pump in retrofit are investigated. Basically, many heat pump concepts can be utilized. To have simple figures, the MVR heat pump, which is a state-of-the-art industrial system, was used to explore their applications to enhance the performance of existing conventional distillation column.

3.2.2 *Working Principle of MVR*

Figure 3.2 shows the scheme of a MVR system, which lifts the temperature of the top vapor to use this as the heat source for the reboiler. The temperature lift can be calculated as follows:

$$\Delta T_{lift} = T_h - T_c = \Delta T_{column} + \Delta T_{HEX} \tag{3.1}$$

where ΔT_{column} is the difference in temperature between the overhead and bottom of the column; and ΔT_{HEX} is the optimized temperature difference over the heat exchanger (typically 10 K).

From the first law of thermodynamics, the amount of heat delivered to the hot reservoir (Q_h) at the higher temperature (T_h) is related to the amount of heat

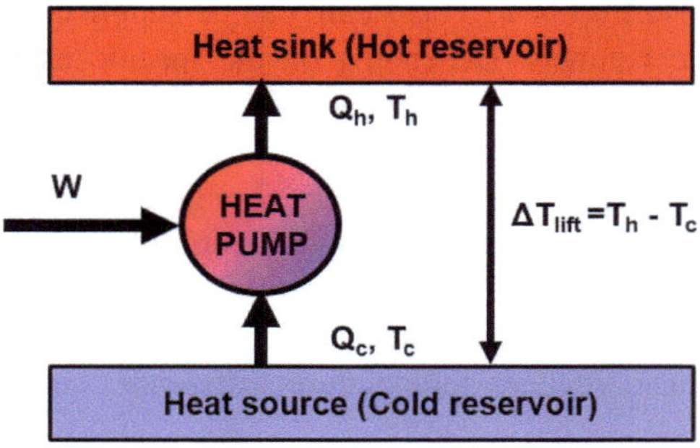

Fig. 3.2 Scheme of a heat pump

extracted (Q_c) from the cold reservoir at the low temperature (T_c) and the external work by the following equation (Bruinsma and Spoelstra, 2010):

$$Q_h = Q_e + W \tag{3.2}$$

The measure of the heat pump performance is the coefficient of performance (*COP*), which is the ratio of the heat rejected at high temperatures to the work input:

$$COP = \frac{Q_h}{W} \tag{3.3}$$

The MVR is strongly recommended if the Q_h/W ratio is more than 10. More evaluations are needed if it is between 5 and 10. When the ratio is lower than 5, a MVR should not be considered for improving the energy efficiency (Pleşu et al., 2014; Long and Lee, 2017). Besides, the inspection of column grand composite curve can be used to evaluate the thermodynamic feasibility of implementing the heat pump to improve the distillation performance, which is discussed in following sections. The upper theoretical value of *COP* obtainable in a heat pump is the COP_c related to the Carnot cycle:

$$COP_c = \frac{T_h}{T_h - T_c} \tag{3.4}$$

3.2.3 MVR Configuration

MVR can be utilized in many processes, such as evaporator, crystallizer, dryer, stripper, and distillation. Most of their applications involved the use of MVR blowers rather than MVR compressors. For a conventional distillation, a MVR can be used directly (Fig. 3.1b), where the heat is transferred directly to reboiler after increasing the temperature. However, with this system, the fouling tendency is not well accepted and auxiliary start up heat is required. In contrast, the sequence can be modified by transferring the heat from the top vapor stream to generate steam, which is then increased the pressure and exchanged the heat in the compressor and reboiler, respectively (Fig. 3.3). This configuration can be considered as an indirect sequence. An indirect configuration can allow fouling and does not require auxiliary start up heat. This steam can be supplied not only to that column but also to other consumers. Depending on the specific example, a direct or indirect configuration can be used. To have simple figures, a direct configuration was used to explore the applications of MVR in other sections.

Note that in the direct configuration, to protect the compressor from the liquid break due to partial condensation of some saturated vapors, the top vapor is led to a compressor after its superheating by using remaining heat in the heat exchanger outlet stream or other hot process streams or utility streams in a superheater. This also reduces temperature difference, compression ratio and compressor power.

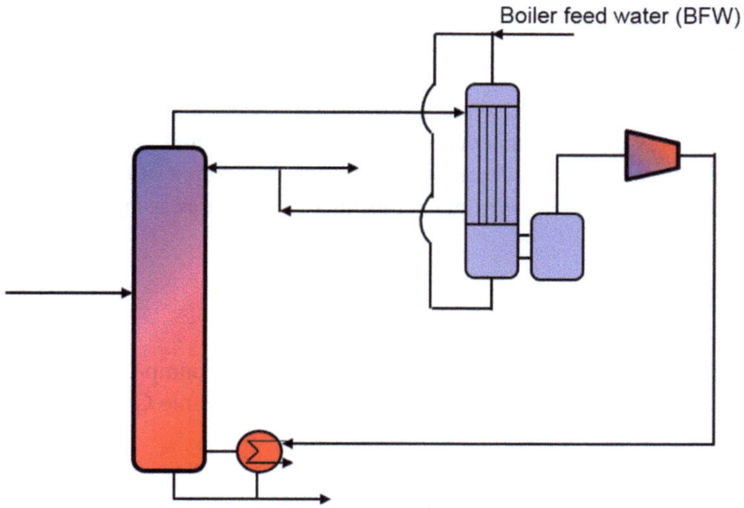

Fig. 3.3 Schematic diagram of the MVR system with indirect way

The same maner can be applied for indirect configuration. Furthermore, the liquid expander can be considered to replace the throttle valve and improve the energy efficiency of the proposed configurations.

3.2.4 *Pros and Cons of MVRs*

In a distillation column, the latent heat of condensation of the overhead is available at the condensers as significant waste heat, which can be recovered to provide a partial/total heating duty at the reboiler or to generate electric power (Chew et al., 2014). A significant amount of energy can be saved by upgrading low temperature waste heat to high temperature using a heat pump. Heat pumps can be used not only in grass-root designs, but also in retrofitting designs because they are easy to introduce and the plant operation is normally simpler than heat integration (Olujić et al., 2006; Long and Lee, 2013b).

A heat pump on the top of the column does not change the vapor and liquid traffic inside the column, whereas a side heat pump can reduce the vapor and liquid traffic on the trays above it and increase the vapor and liquid traffic below it. This will decrease the area utilization above the side heat pump, that is, it can increase the capacity of the distillation column when it has the maximum flooding in the top section.

On the other hand, the high capital expenditure needed for compressors make them industrially viable only for high-capacity end-of-train (practically binary) separations of substances with similar boiling points, which require minimal compressor/compression cost (Suphanit, 2011). For wide-boiling mixtures, the use of a side heat pump or a side reboiler can overcome the problems related to high compressor cost because of the high temperature difference between the top and bottom of the column (Lynd and Grethlein, 1986; Long and Lee, 2017); however, it may cause more difficulties in controlling these complex systems.

When a heat pump is used, engineers should consider the reactions that take place as the vapor pressure and temperature are increased (Long and Lee, 2014). Engineers also should be aware of the decomposition or polymerization of components, which are sensitive to temperature when the temperature is increased using a heat pump.

3.2.5 Case Study

The below example illustrates the benefits of using heat pump to retrofit the distillation. The existing conventional distillation is used in whole Chap. 3.

3.2.5.1 Existing Conventional Distillation

A deisobutanizer, possessing 92 theoretical trays (Amminudin and Smith, 2001; Long and Lee, 2013b) was designed and operated at 6.5 bar (Fig. 3.4) because commercial isobutane can be condensed with cooling water under this pressure (Klemola and Ilme, 1996). The maximum flooding of the modeled columns was determined using a rating mode simulated using the internal specifications of the columns, such as the type of trays, column diameter, tray spacing, and number of passes. The simulations were run using the simulator, ASPEN HYSYS V8.4. The Peng-Robinson equation of state was used to predict the vapor-liquid equilibria of these simulations (Aspen Technology, 2009). The base case simulation model showed that the energy consumption of the deisobutanizer was 6544 KW.

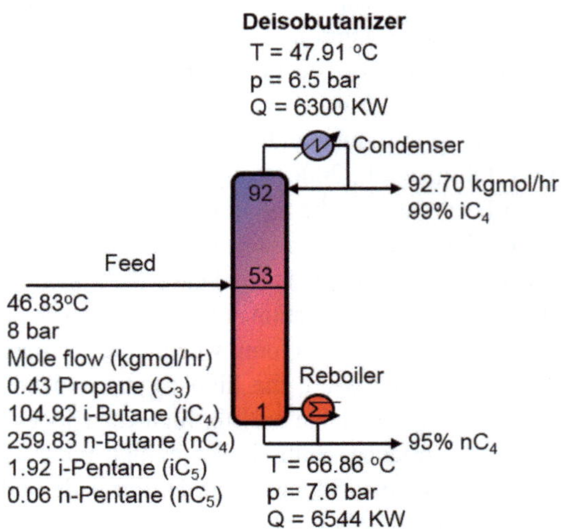

Fig. 3.4 Simplified flow sheet illustrating the conventional column – deisobutanizer

3.2.5.2 Heat Pump Assisted Distillation

Distillation columns are thermodynamic systems comprising a heat source (condenser) and heat sink (reboiler) that can be represented graphically on a temperature–enthalpy diagram known as the column grand composite curve (CGCC) (Rivera-Ortega, 1997). Figure 3.5 shows the CGCC for the deisobutanizer. An inspection of the CGCC confirmed that the column energy efficiency can be improved using a heat pump operating between the top and bottom with relatively small differences in temperature (Fig. 3.5a). The distance between the points on the curve and the vertical axis indicates the heat load required for each stage (Fig. 3.5b). When the heat load is high, there is a strong likelihood of improving the energetic efficiency of the system.

Vapor from the top of the distillation column can be compressed to a certain pressure (increasing its vapor temperature) and condensed in the column's reboiler (Gadalla, 2009). This allows the heat of the condensing vapor to assist in vaporization at the bottom of the column via a heat exchanger. The top outlet stream must be cooled further before being divided into two streams: one that is recycled back to the column as reflux, and the other that is the final top product (Fig. 3.6). The pressure of the outlet compressor is adjusted to obtain the minimum approach in the heat exchanger of 10 °C. Compared to the conventional column sequence, the use of a top vapor recompression heat pump can reduce the condenser duty, reboiler duty and operating costs by 72.6, 80.8 and 67.2%, respectively. Table 3.1 lists the utility cost data (Turton et al., 2012). The CO_2 emission reduction associated with the lower energy requirement is another important benefit of using the proposed sequence in the retrofit of existing deisobutanizer. The method reported by Gadalla et al. (2005) was used to estimate the CO_2 emissions. The study results showed that CO_2 emissions can be reduced dramatically when a part of the process heat is recovered for use instead of the primary steam. In particular, the annual amount of CO_2 emissions is reduced up to 55.5%. The heat pump assisted distillation results in smaller CO_2 emission reduction than the operating cost saving (67.2%).

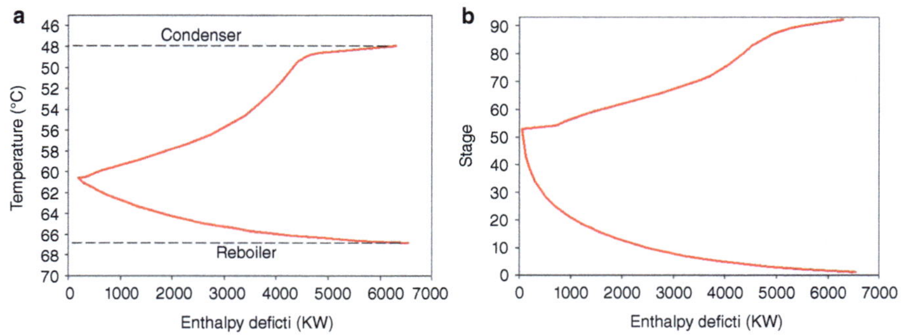

Fig. 3.5 CGCC for a conventional column

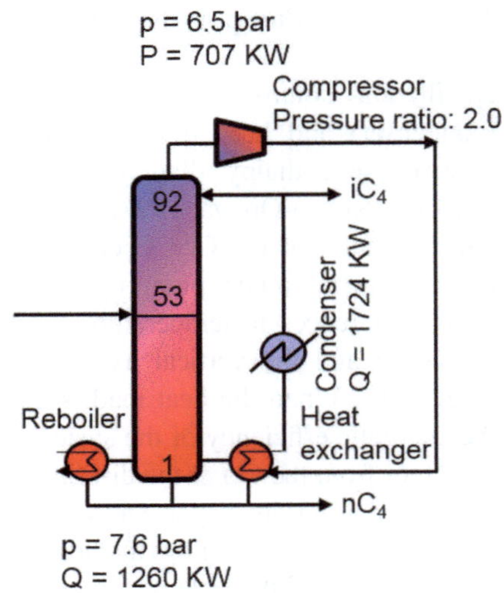

Fig. 3.6 Simplified flow sheet illustrating the retrofitted mechanical vapor recompression heat pump

Table 3.1 Utilities cost data

Utility	Price ($/GJ)
Cooling water	0.35
Steam (LP)	13.28
Steam (MP)	14.19
Steam (HP)	17.70
Electricity	16.80
Refrigeration (moderately low temperature)	4.43
Refrigeration (low temperature)	7.89

3.3 Self-Heat Recuperation Technology

3.3.1 Introduction

In the distillation process, heat is supplied to the feed heater and reboiler, and the overhead stream is cooled in a condenser. Most of the heat supplied to the reboiler in the conventional distillation process is discarded in the overhead condenser (Matsuda et al., 2011). Heat pumps in the distillation allow the heat of condensation released at the condenser to be used for evaporation in the reboiler (Bruinsma and Spoelstra, 2010).On the other hand, in heat pumps, only the heat recovery duty to the reboiler in the distillation column is considered; the heat during preheating is not well recognized (Matsuda et al., 2011) and only latent heat is utilized while the sensible heat is neglected (Long and Lee, 2013b).

Recently, (Kansha et al., 2010) developed a novel integrated process module based on self-heat recuperation and facilitated heat circulation for use in distillation processes. By dividing the distillation processes into two modules, that is, heat circulation and distillation modules, the heat of condensation and the cooling

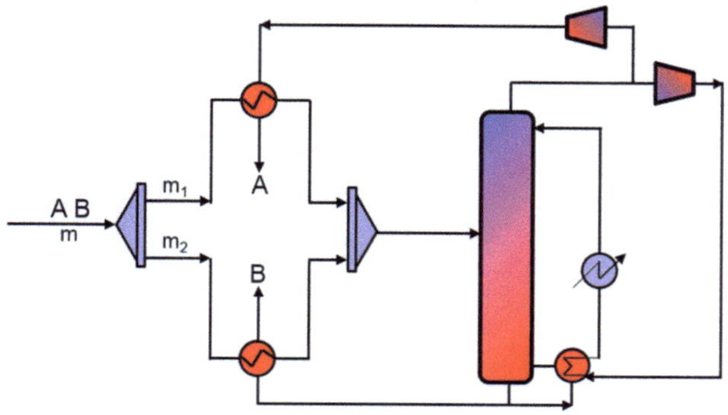

Fig. 3.7 Schematic diagram of the distillation with self-heat recuperation technology

heat were recovered by compressors and exchanged with the heat of vaporization and the heat for heating through the use of heat exchangers in each module, without heat addition. In this technology, two compressors were installed in the overhead vapor lines, as shown in Fig. 3.7. A compressor treats the reflux stream and the other compressor treats the overhead stream (Matsuda et al., 2011).This self--heat recuperation technology facilitates the recirculation of both the latent and sensible heat in a process, and helps reduce the energy consumption of that process using compressors and self-heat exchangers (Kansha et al., 2009; Matsuda et al., 2011; Kansha et al., 2013).

3.3.2 *Case Study*

This example is distillation retrofit using integrated distillation based on self-heat recuperation. Self-heat recuperation technology utilizes both latent and sensible heat in the process using a compressor. In self-heat recuperation technology, the heat of the streams is recovered using compressors, and the recovered heat of the pressurized stream is supplied to the cold stream via a heat exchanger. To retrofit the deisobutanizer, one new compressor and three heat exchangers are installed. The compressor treats all the overhead vapor, and the compressed vapor is divided into the reboiler and feed heater. The feed stream is divided into two parallel streams, which are exchanged with the overhead and bottom product streams to maximize the heat recovery duty. Correct adjustment of the feed split ratio is important for maximizing the heat recovery and compressor power. Note that the split is defined as the ratio between the flow that is heat exchanged with the bottom product stream to the total flow (m_2/m). An optimal feed split ratio of 0.72 was found, which decreased the annual operating costs by 69.4% compared to the conventional sequence (Fig. 3.8). Figure 3.9 presents a simplified flow sheet outlining integrated distillation process based on self-heat recuperation. The simulation showed that the top vapor stream increases when the feed receives some heat, that is, more heat can be recuperated. In addition, the reduced boil-up flow decreases the flooding value in the bottom section.

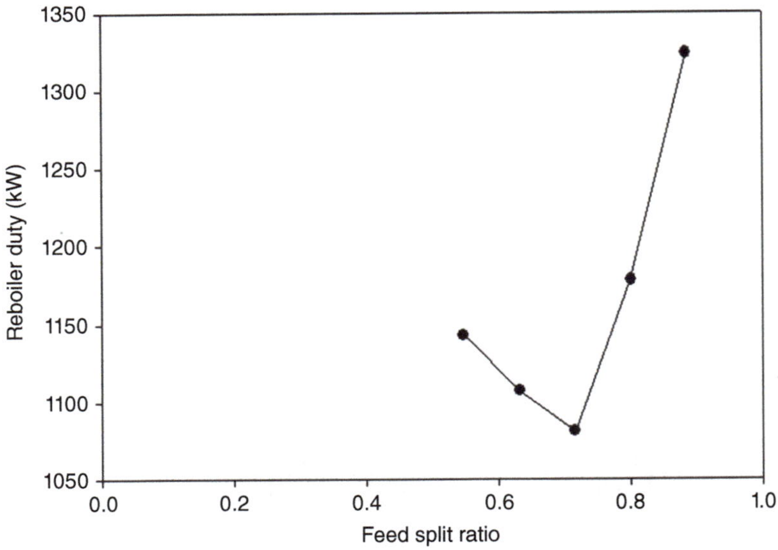

Fig. 3.8 Effect of the feed split ratio on the reboiler duty

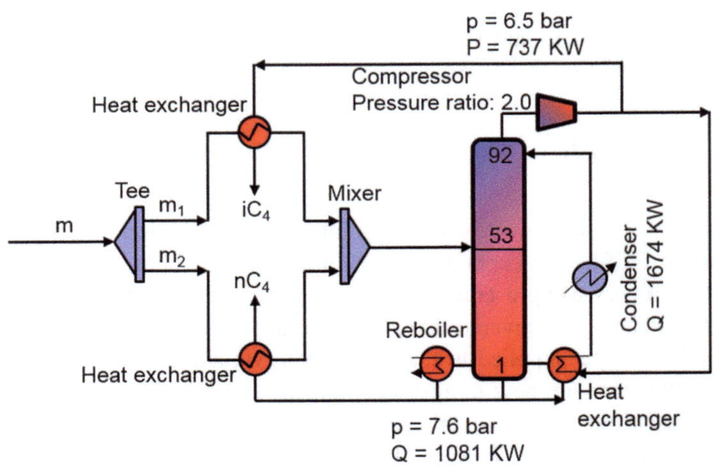

Fig. 3.9 Simplified flow sheet illustrating the distillation column with the heat circulation module based on self-heat recuperation

Instead of dividing into two streams, the compressed vapor is condensed in the column's reboiler, which then supplies heat to the feed stream. The feed stream is also exchanged with the bottom product streams. Figure 3.10 presents a simplified flow sheet showing the improved self-heat- recuperative distillation process, which has two exchangers in series (HES configuration). This arrangement reduces the annual operating cost by 79.0% compared to the conventional column sequence.

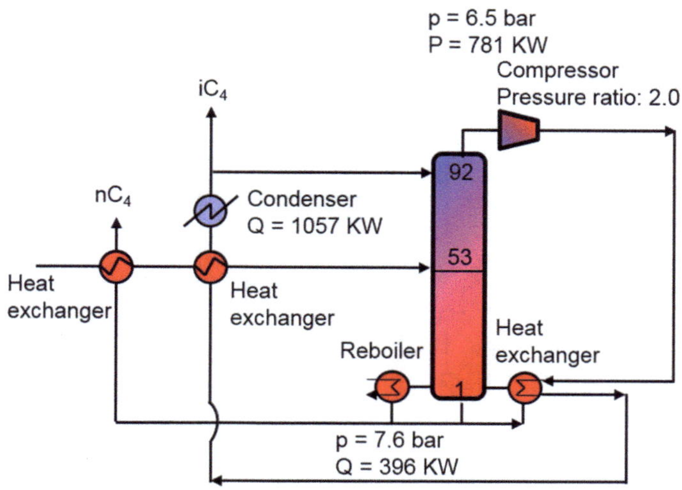

Fig. 3.10 Simplified flow sheet illustrating the proposed integrated distillation process (HES)

To reduce the energy requirements in the distillation unit using a heat recovery technique, the heating and cooling functions are normally integrated based on heat exchange between the feed and product streams. On the other hand, temperature differences between the two streams are required to exchange the heat according to the second law of thermodynamics. To maximize the heat recovery duty, the feed stream is divided into two parallel streams (HEP configuration). Vapor from the top of the distillation column is compressed and condensed in the column's reboiler. The top outlet stream then exchanges that heat with the feed stream and is cooled further before being divided into two streams: one that is recycled back into the column as reflux, and the other that is the final top product (Fig. 3.11). The feed split ratio is varied to maximize the heat recovery duty and minimize the compressor power. The optimal feed split ratio was found to be 0.34 (Fig. 3.12). As a result, all the heat is recirculated in the process without the need for external heat, and only electric power for the compressors is needed. Savings in the condenser energy requirement, reboiler energy requirement and operating cost of up to 64.4%, 100.0% and 52.8%, respectively, compared to a classical heat pump system, and 90.2%, 100.0% and 84.5%, respectively compared to the conventional sequence could be achieved, as shown in Table 3.2.

This methodology can be applied when the system satisfies the following two conditions: (1) there is a source of heat; and (2) there is a temperature difference between the two streams sufficient to exchange heat according to the second law of thermodynamics. More heat can be recovered as the feed temperature approaches its bubble point.

Figure 3.13 presents the flooding profiles of all the sequences. In the conventional column and heat pump system, the vapor and liquid traffic are high in the bottom section, which results in a higher flooding profile. In integrated distillation based on self-heat recuperation that facilitates recirculation, the feed is vaporized

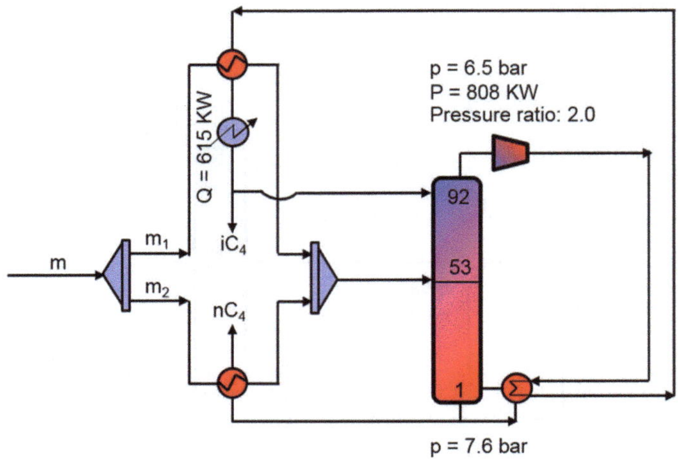

Fig. 3.11 Simplified flow sheet illustrating the proposed integrated distillation process (HEP)

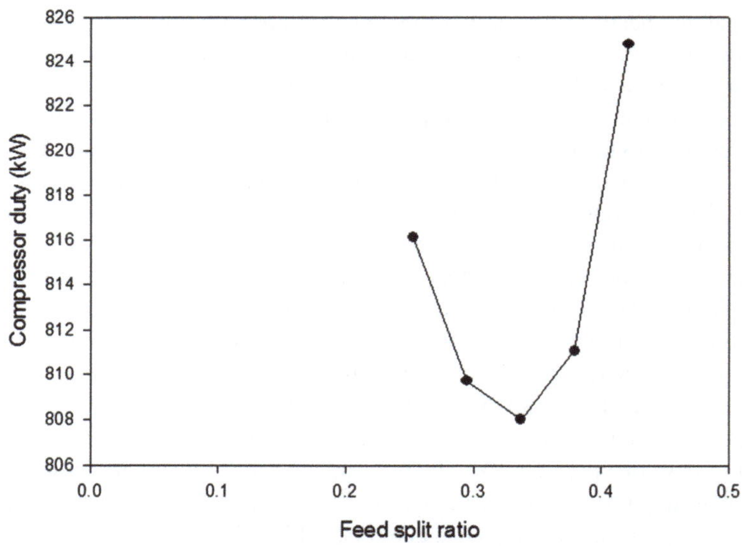

Fig. 3.12 Effect of the feed split ratio on the compressor duty

partially. As a result, the vapor and liquid traffic is increased in the top section but decreased in the bottom section.

This is a promising technology when the problem is associated with binary mixture separation. From the existing column, new heat exchangers and compressors can be installed while the number of trays and the diameter of the column are fixed. More pipe work will also be needed for the equipment link. From a technical point of view, those studies can be accomplished in an easy manner and short modification time.

Table 3.2 Summary of relative performance for various arrangements

Structural alternative	Conventional column	Heat pump	Distillation based on self-heat recuperation	Proposed integrated distillation process (HES)	Proposed integrated distillation process (HEP)
Investment costs (US $)	–	2,802,995	3,046,868	3,314,788	3,400,549
Energy requirement saving in the condenser (%)	–	72.6	73.4	83.2	90.2
Energy requirement saving in the reboiler (%)	–	80.8	83.5	94.0	100.0
Annual operating cost saving (%)	–	67.2	69.4	79.0	84.5
Annual operating cost saving (US $/year)	–	1,863,574	1,922,391	2,189,115	2,343,431
Payback period (month)	–	18.1	19.0	18.2	17.4

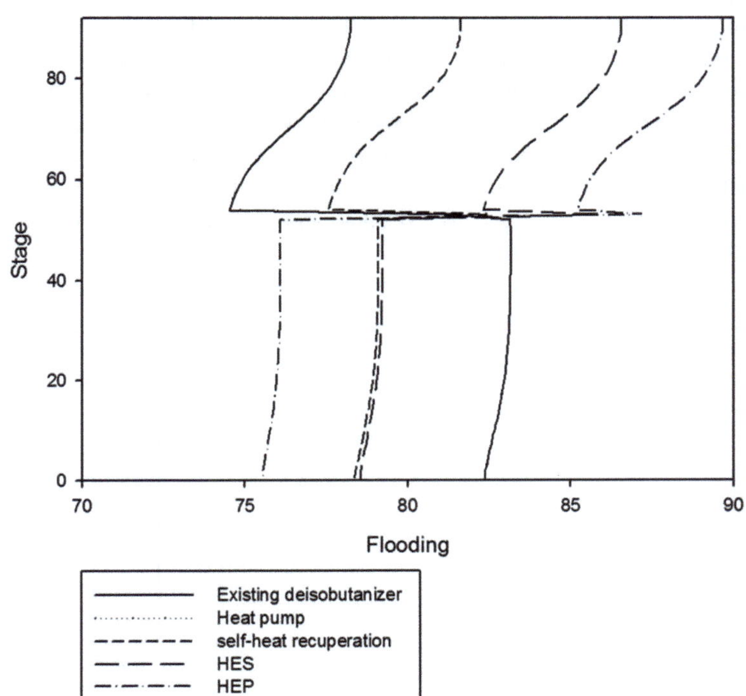

Fig. 3.13 Flooding profile

3.4 Feed Thermal Condition

3.4.1 Introduction

The thermal condition of the feed has a strong influence (Liu and Jobson, 2004) on the processing capacity of an existing distillation column. This is particularly true for columns performing difficult separations. This discussion focuses on distillation columns with a single feed and only two products, but these ideas can also be applied to more complex columns. The thermal conditions of the feed can be altered using a feed preheater or precooler (Fig. 3.14). The feed condition affects the amount of vaporization required in the reboiler and the amount of condensing required, that is,. the heating and cooling duties (Jobson, 2014). According to the author, the total heating and cooling duties tend to be lowest when the feed enters as a saturated liquid. Therefore, from the point of view of duty only, a saturated liquid feed is most energy efficient. In addition, the hot utility suitable for heating the feed maybe cooler and cheaper than that used in the reboiler, as the bubble point and dew point temperatures of the feed are generally lower than that in the reboiler (Jobson, 2014). This should be considered in a retrofit to reduce the energy requirement.

Furthermore, because the thermal condition of the feed entering a distillation column have a direct impact on the flow rate of vapor and liquid within the column (Jobson, 2014), changing the thermal condition can be applied in retrofit to increase the capacity. In particular, a feed preheater can reduce the vapor and liquid traffic in a stripping section, and increase the vapor and liquid traffic in a rectifying section. Similarly, a feed precooler can reduce the vapor and liquid traffic in a rectifying section, and increase the vapor and liquid traffic in stripping section.

To reduce the reboiler duty, the feed can exchange heat with the bottom product or top vapor or with any other low-grade heat sources available (Fig. 3.15a and 3.15b). Furthermore, splitting the feed into two streams and preheating only one fraction of the feed stream makes it is possible to improve the feed preheat efficiency (Fig. 3.15c and 3.15d) (Wankat and Kessler, 1993; Fidkowski and Agrawal, 1995; Soave and Feliu, 2002).

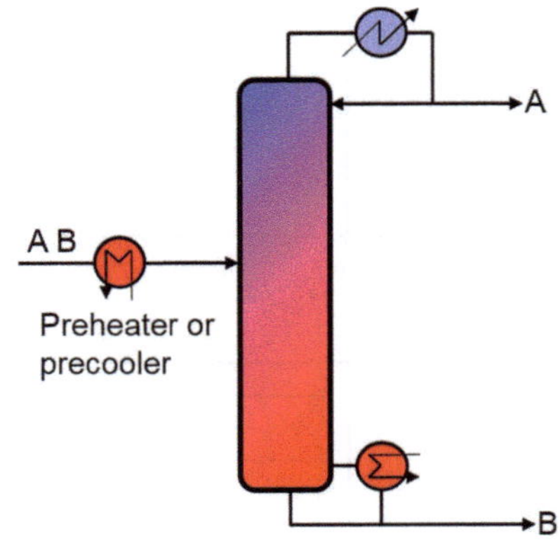

Fig. 3.14 Simplified flow sheet illustrating the conventional column with a preheater or a precooler

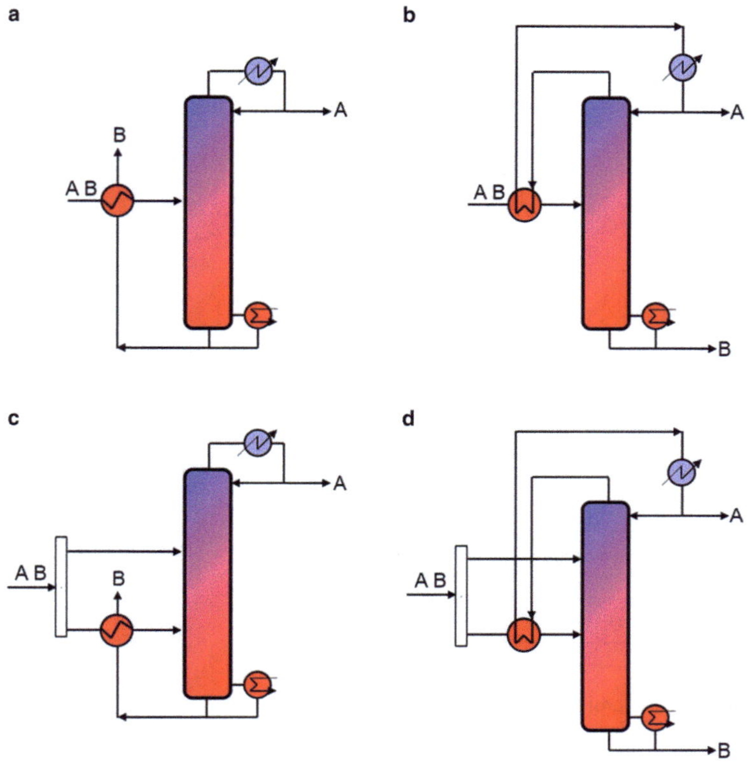

Fig. 3.15 Simplified flow sheet illustrating the (a) preheating using bottom product heat, (b) preheating using the top vapor heat, (c) feed split and preheating using the bottom product heat and (d) feed split and preheating using the top vapor heat configurations

3.4.2 Case Study

The above example is also used to illustrate how the feed preheat affects the total heating duty, flooding and capacity. The shape of the flooding profile can be used to identify opportunities to adjust the feed condition to increase the capacity of a distillation column. In particular, as shown in Fig. 3.16, the flooding in the bottom of the column is higher than that in the top of the column. This means that the feed thermal condition may not be appropriate from the point of view of capacity. To increase the capacity, first the vapor and liquid traffic in the column require re-arranging. In particular, preheating the feed using a preheater can be employed, which impacts directly on the flow rate of vapor and liquid within the column, as shown in Fig. 3.16.

Figure 3.17 shows the influence of the feed preheat duty on the total duty, reboiler duty, and condenser duty. The heat duty in the reboiler decreases with increasing heat duty in the feed preheater. On the other hand, the total duty and condenser duty increase with increasing duty in the feed preheater. In addition, the maximum

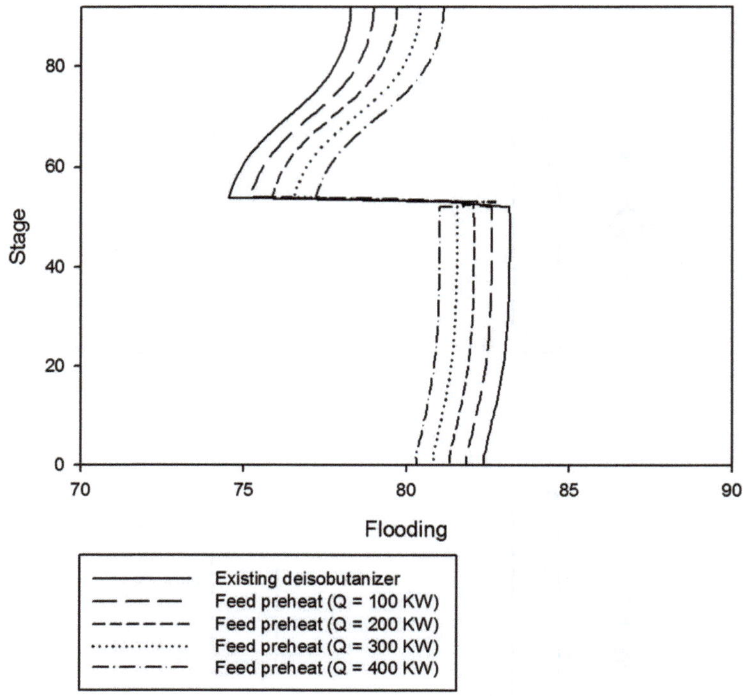

Fig. 3.16 Flooding profile

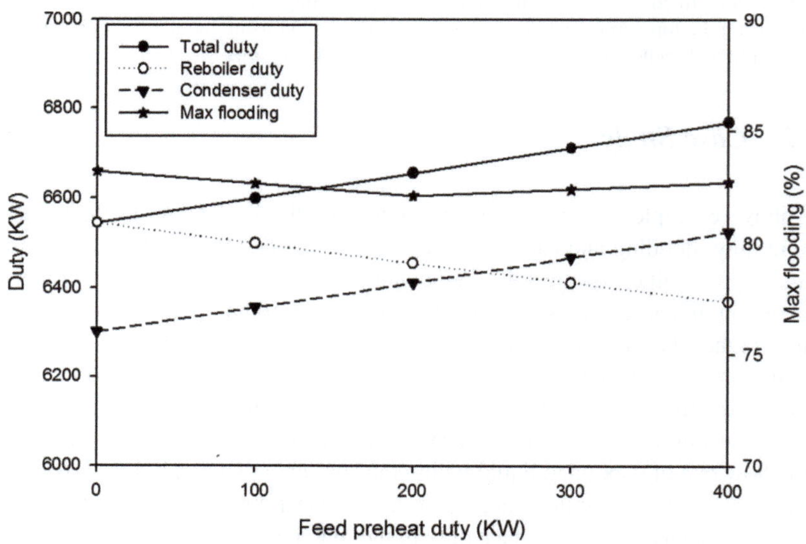

Fig. 3.17 Influence of the feed preheat duty on the total duty, reboiler duty, condenser duty and max flooding

flooding first decreases with increasing feed preheat duty. In this case, however, when feed preheat duty continues to increase, the feed temperature is between its bubble point and dew point temperatures, which causes a peak in flooding profile at the feed stage (Fig. 16). This is why although the flooding profile in the bottom section decreases with increasing feed pre heat duty, the maximum flooding is increased after the saturated liquid point (Fig. 3.17). Note that when feed stage shows a peak in flooding profile, it may not follow that flooding is likely on the feed stage because the capacity is increased (Liu and Jobson, 2004). The reason for this is that the tray spacing on this tray is normally increased to accommodate the liquid distributors, pipes, and an access path. Therefore, the peak in the feed stage can be ignored. Even it is serious, splitting the feed into two streams and preheating only one stream can remove the flooding problem in the feed stage.

3.5 Side Reboiler and Side Condenser

3.5.1 Introduction

In a distillation column, the bottom has the highest temperature, which requires the most expensive heat source to operate the column. An alternative is the use of an intermediate reboiler located higher up the distillation column, where the temperature levels are lower – hence a lower-temperature, less expensive heat source can be used (Fig. 3.18a) (Kiss, 2013). The heat duty in the reboiler decreases with increasing heat duty in the side reboiler. On the other hand, adding a side reboiler may require a higher total heat input and removal (Long and Lee, 2013c).

Similarly, a side condenser can reduce the amount of heat rejected by the condenser (at the lowest temperature in the column) and provide a potentially useful heat source at a more moderate temperature (Fig. 3.18b) (Jobson, 2014). This can be attractive when refrigeration is used to condense the top vapor stream. The condenser duty decreases with increasing side condenser duty; however, the withdrawal a vapor stream may cause operation problems. Therefore, it is often more practical to withdraw a liquid stream, subcool it and return it to the column; upon entering the

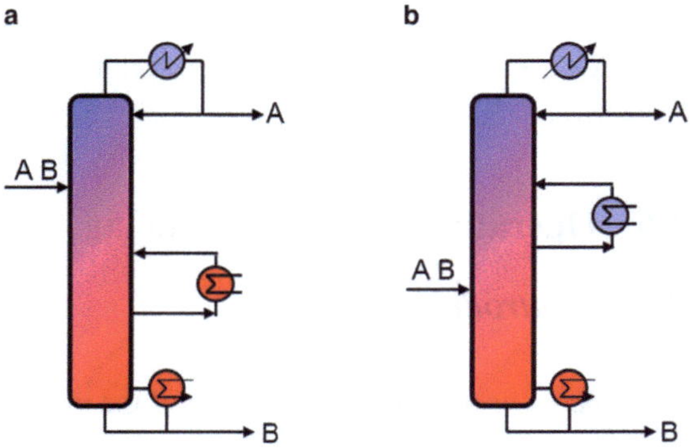

Fig. 3.18 Simplified flow sheet illustrating the conventional column with (a) a side reboiler and (b) a side condenser

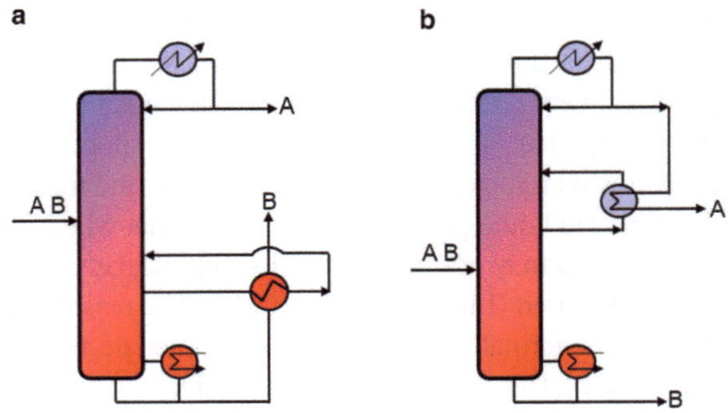

Fig. 3.19 Simplified flow sheet illustrating the conventional column with (a) a side reboiler using bottom product heat and (b) a side condenser using distillate product as a cold stream

column, the subcooled liquid causes some of the vapor to condense and flow down the column, thereby increasing the reflux below the return stage (Jobson, 2014).

The use of a side reboiler or side cooler can also allow the throughput of a distillation column to increase (Liu and Jobson, 2004; Long and Lee, 2013c). Side reboilers and side coolers can be used to alter the vapor and liquid flows within a column section to utilize the area available more uniformly. In particular, a side reboiler will reduce the vapor and liquid traffic below the tray, where the side reboiler is placed, and increase the vapor and liquid traffic above the tray. In a similar manner, a side cooler will reduce the vapor and liquid traffic on the trays above it and increase the vapor and liquid traffic below it. Therefore, this process modification increases the uniformity of area utilization throughout the column section.

The bottom liquid product heat can be used to minimize the energy consumption when using a side reboiler (Fig. 3.19a). This technique is known as inter-reboiling (Manley, 1998; Long and Lee, 2012). This heat effect is particularly significant when the feed contains significant amounts of heavy components. The temperature profile should be checked to choose the side reboiler stage with a much smaller temperature than the bottom temperature. The stages of the side reboiler draw and return can be different. To have easy operation, the return stage with a lower pressure is above the draw stage, which has a higher pressure. This configuration can be used effectively for both reducing energy requirement and increasing the capacity. In a similar manner, the distillate stream can reduce the condenser duty when it plays as a cold stream in a side condenser (Fig. 3.19b). This technique referred to as inter-condensing has more benefits when that distillate stream enters another thermal separation unit in downstream.

3.6 Operating Pressure Changing

3.6.1 Introduction

The column operating pressure is one of the most important distillation design parameters; it affects the temperatures at which heating and cooling are required, the type of heating and cooling utility required, as well as the heating and cooling duties

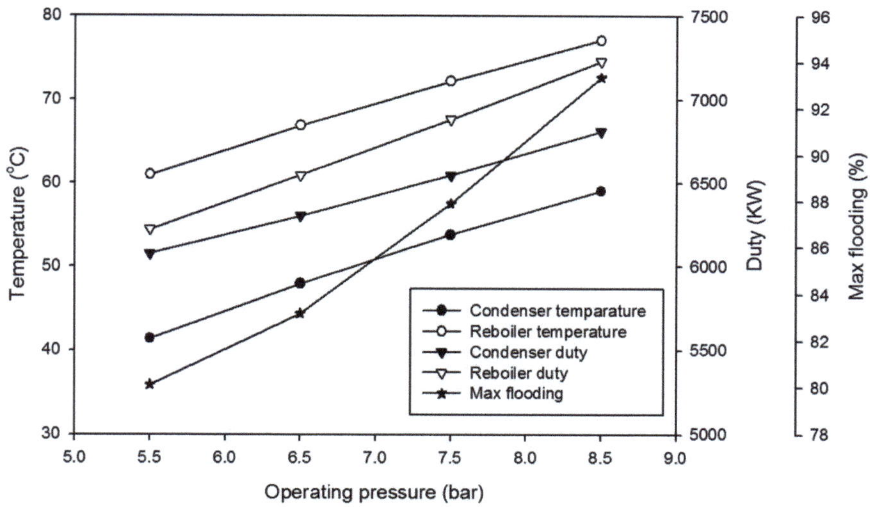

Fig. 3.20 Influence of the operating pressure on the condenser temperature, reboiler temperature, condenser duty, reboiler duty and max flooding

(Jobson, 2014). In particular, a higher pressure increases the condenser and reboiler temperature, as shown in Fig. 3.20. This requires a more expensive heating source while the cooling cost can be reduced. There is a reduction in the relative volatility as pressure is increased, which increases the separation difficulty. This increases the reflux, stage requirements and reboiler and condenser duties (Kister, 1992), which results in higher investment and operating cost to achieve the same target.

Figure 3.20 also shows the influence of the operating pressure on the condenser duty, reboiler duty and maximum flooding of deisobutanizer. A decrease in operating pressure can decrease the reboiler and condenser duty. In addition, flooding is reduced when the operating pressure is decreased, which allows an increase in the capacity of the distillation column.

The column is normally operated at atmospheric pressure unless there are some reasons, such as increasing the operating pressure to avoid refrigeration, decreasing the operating pressure of the vacuum condition to avoid the degradation of thermally sensitive materials, and changing the operating pressure to create opportunities for heat integration (Jobson, 2014). Note that when increasing the operating pressure in the separation of very light components, it needs to check whether the operating pressure is near the critical pressure of the key components or not because the density difference between the liquid and vapor phases becomes too small. A common heuristic is to set the pressure at half the critical pressure (Luyben, 2016). The resulting condenser pressure for these systems requires refrigeration or vapor recompression.

In addition, it can facilitate separation when a mixture forms an azeotrope, which is pressure sensitive. As shown in Fig. 3.21, with increasing operating pressure, the azeotrope moves to left direction, which has opportunity to increase the recovery of the heavy product in the bottom stream (main product) while the top stream is recycled to the tank for another purpose. This will also increase

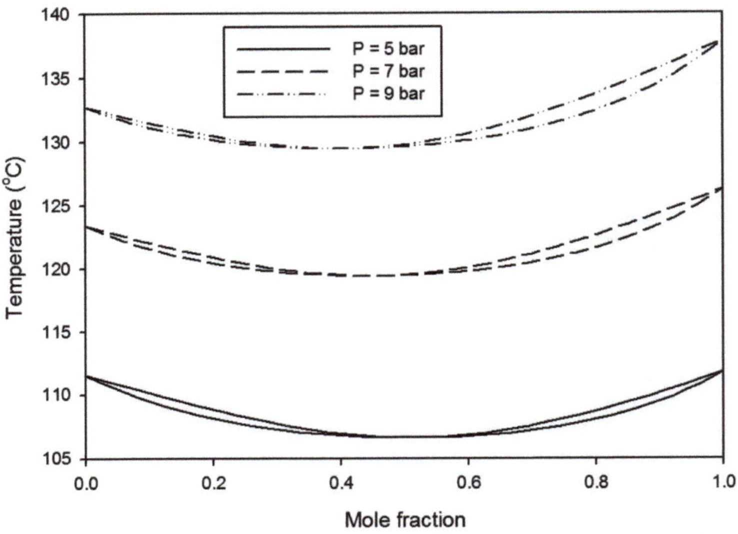

Fig. 3.21 Influence of the operating pressure on the composition of azeotrope

the separation ease on the right side of the azeotrope, which may reduce the energy requirement and flooding. In addition, in hot country or in summer, the column operating pressure would be increased to use a cheap cooling source, such as water. When the product is in the vapor phase, which requires a higher pressure to meet the downstream requirement, the column is operated at a higher pressure to eliminate the compressor.

In a retrofit design, design engineers are often under pressure to increase the capacity of an existing distillation column. Among the operating parameters, pressure is very important (Liu and Jobson, 1999), and has a complex influence on the distillation performance; it affects the relative volatility between the components, molar vapor flow rates, flooding limitations and vapor density.

Many studies reported that increasing the pressure can decrease the diameter of a distillation column for a given separation (Liu and Jobson, 1999). Kister (1992) reported that one of the favorable effects when increasing the column pressure is that it increases the vapor density and vapor handling capacity. This leads to a major decrease in the column diameter and capital cost under vacuum, as well as to smaller decreases to pressures of 3.4–10.2 bar. Humphrey and Keller (1997) suggested that at higher pressures and temperatures, the vapor pressures increase and the viscosities decrease, which means greater efficiency and vapor capacity, as well as a lower column diameter.

Other studies have reported that decreasing the pressure can increase the capacity of a distillation column (Liu and Jobson, 1999). Capps (1993) found that for high-pressure systems, a decrease in the operating pressure allows an increase in feed rate. This occurs because of a decrease in the relative volatility and flooding capacity. Castillo and Dhole (1995) examined the ethylene cold-end process and suggested that the capacity of the columns in the process could be increased by decreasing the operating pressure of the demethanizer from 35 bar to 31 bar.

Therefore, the effects of pressure on the capacity and recovery of an existing distillation column should be checked for specific cases.

3.7 Multi-Effect Distillation

3.7.1 Introduction

Frequently, more than 95% of the energy added to the reboiler of a distillation column is removed by the cooling water at the top of the column (Cheng and Luyben, 1985). Many energy-saving configurations have been known for a long time, but only recently have they become attractive because of the drastic increase in energy cost. The approaches to effect energy conservation in distillation by utilizing the heat from the top vapor stream generally fall into the following categories:

1. Feed preheat
2. Multi-effect distillation
3. Heat pump assisted distillation

As discussed in above sections, the concept of a heat pump assisted distillation is to use the overhead vapor as a heat source for the reboiler. The adverse temperature difference is overcome by adding compression work to the vapor to raise its pressure and condensing temperature. Distillation schemes with heat pumps require considerable compression work when the temperature difference to be overcome is large. Therefore, they are not often used high relative volatility systems.

Multi-effect distillation means that the column pressures are adjusted such that the cooling (energy removal) in one column can be used for heating (energy input) in another column (Engelien and Skogestad, 2005). This method of heat integration reduces the overall heat demand of a distillation process because the heat supplied to the high-pressure (HP) column is also used for distillation in the low-pressure (LP) column (Agrawal, 2000).

Figure 3.22a presents the configuration of a multi-effect distillation configuration for a binary mixture with two distillation columns operating at two different operating pressures. The feed to the process is split into two streams and fed to two columns. The pressure in the HP column is selected to allow the heat rejected from the condenser of the HP column to be transferred to the reboiler of the LP column in a heat exchanger (Jobson, 2014). With this configuration, the requirement for hot and cold utilities can be reduced substantially compared to that in a single column carrying out the same separation. This sequence is quite attractive when a large capacity, such as 50% is increased, which may require the installation of new column to take care the extra feed.

When the temperature is quite high and there is a sudden increase in the bottom temperature, while the intermediate point in the column has a much lower

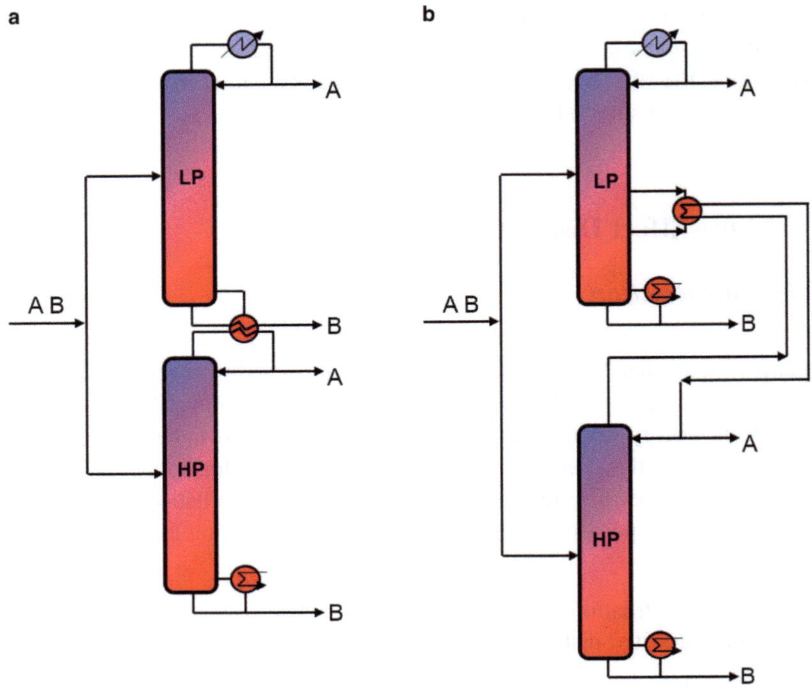

Fig. 3.22 Schematic diagram of the (a) multi-effect distillation and (b) multi-effect distillation with a side reboiler

temperature than the bottom temperature, multi-effect distillation with a side reboiler configuration (Fig. 3.22b) can be considered. Note that the relative volatility changes with pressure that may make some difference (Robinson and Gilliland, 1950).

3.7.2 *Case Study*

3.7.2.1 Existing Conventional Distillation

The deisobutanizer is considered an existing conventional distillation. The column is under the assumption that the capacity must increase by 50% over the standard operation capacity and is already operating with the highest performance internals.

3.7.2.2 Debottlenecking

With an increase in production capacity of 50%, the existing column has a bottleneck problem. Therefore, a new column is installed to take care of the extra flow rate (Fig. 3.23a). To reduce the energy requirement, the multi-effect distillation

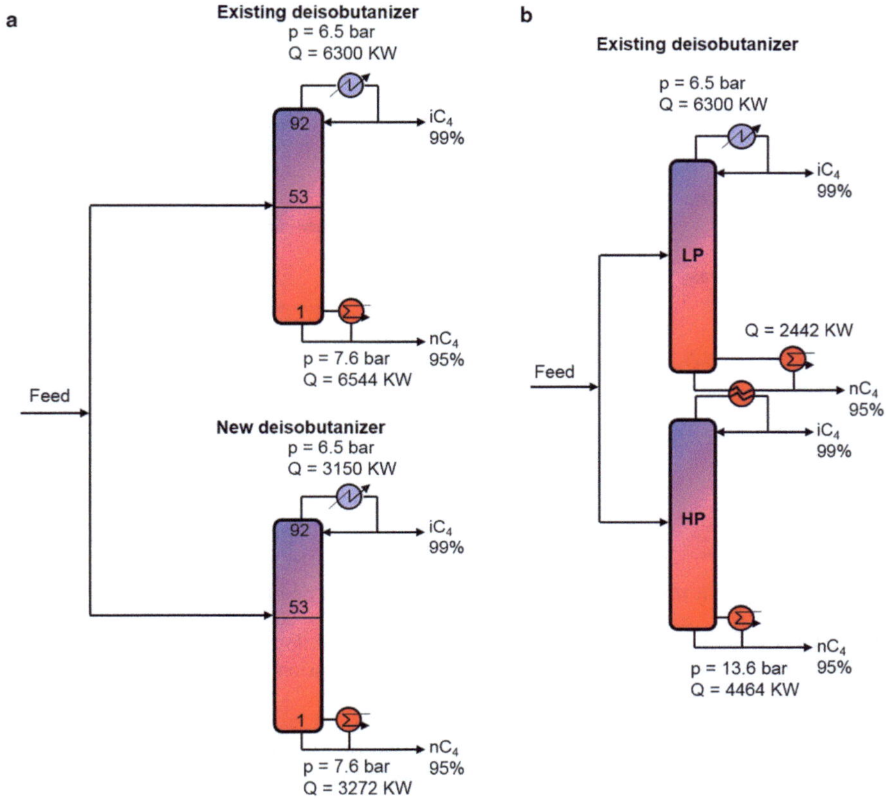

Fig. 3.23 Simplified flow sheet illustrating the (a) conventional column sequences including a new column added in parallel with the existing column; (b) multi-effect distillation

configuration can be considered with the new column operating at high pressure (HP column) to allow the heat rejected from its condenser to boil up the bottom stream of the existing column (LP column) (Fig. 3.23b). Therefore, the total reboiler duty saved can be up to 29.6%. For further improvement, feed preheat, heat pump assisted distillation, or self-heat recuperation technique can be employed simultaneously with multi-effect distillation.

3.8 Cyclic Distillation

3.8.1 Introduction

Cyclic distillation is an intensified unit that has attracted increasing attention recently. It enhances the separation efficiency through the pseudo-steady-state operation based on separate phase movement (SPM), which can save up to 50% of the energy requirement (Gaska and Cannon, 1961; Maleta et al., 2011). The cyclic

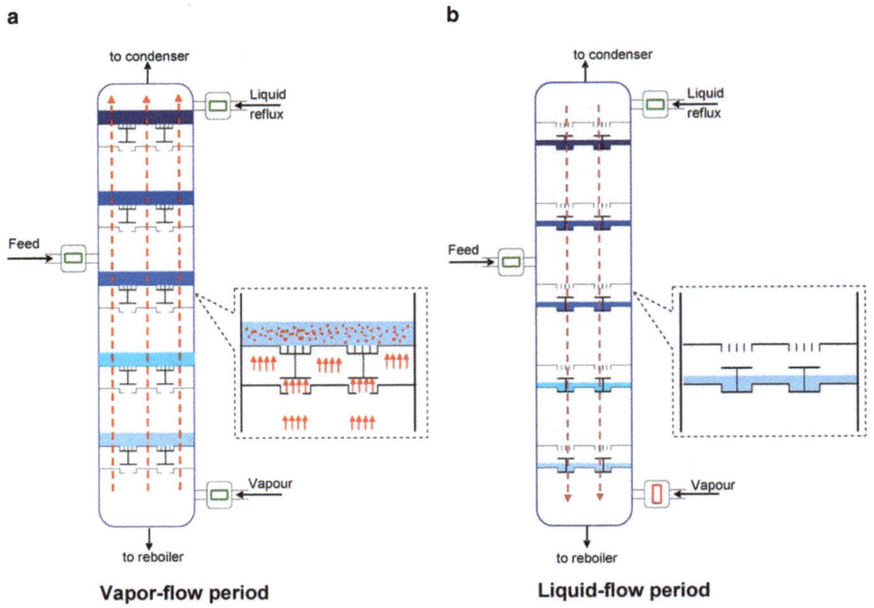

Fig. 3.24 Cyclic distillation works by alternating (a) a vapor-flow period with (b) a liquid-flow period (right). (Reprinted from Maleta et al. 2015, with permission from John Wiley and Sons)

operating mode consists of two parts: (1) a vapor flow period, when vapor flows upward through the column, while liquid remains stationary on each plate, and (2) a liquid-flow period, when the vapor flow is stopped, reflux and feed are supplied to the column, while the liquid holdup is dropped from each tray to the tray below (Maleta et al., 2011; Patrut et al., 2014; Maleta et al. 2015) (Fig. 3.24). This operating mode can have some benefits, such as lower capital expenditure due to the smaller number of trays required, reduced energy requirements and significant savings in operating cost, increased quality of the products, due to higher separation efficiency, higher yield and equipment productivity compared to conventional distillation (Maleta et al., 2011).

Kiss listed several limitations of cyclic distillation, such as the application to vacuum distillation appears rather difficult; the performance enhancement depends critically on complete separation between the liquid and vapor flow periods; and it is difficult to achieve when the column has more than ten simple trays (Kiss, 2013). Maleta et al. (2011) proposed sluice-champer trays to avoid the limitation associated with the number of trays (Fig. 3.25).

Maleta et al. (2011) modeled a cyclic distillation column implemented on an industrial scale in the food industry (Lipnitsky Alcohol Plant, Ukraine). The feed stream with a concentration of alcohol of 10 vol.% is supplied on the top tray of the 15-trays cyclic distillation column (stripping column), while direct steam injection is used in the bottom of the column. They reported that the separation efficiency in the cyclic distillation column is much higher than the conventional

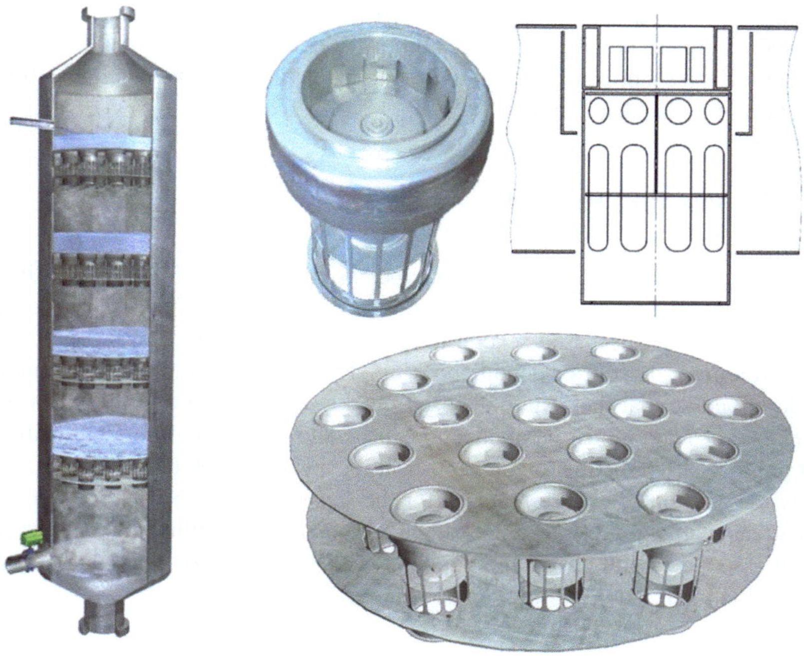

Fig. 3.25 Cross-section of a cyclic distillation and specific internals. (Reprinted from Maleta et al. 2015, with permission from John Wiley and Sons)

bubble cap trays column (56 trays), exceeding 200–300%. Maleta et al. (2015) first reported the performance of a pilot-scale distillation column for ethanol-water separation, which was operated in a cyclic mode. A comparative study was made between a pilot-scale cyclic distillation column and an existing industrial beer column used to concentrate ethanol. Using specially designed trays that truly allow separate phase movement, the practical operation confirmed that 2.6 times fewer trays and energy savings of approximately 30% are possible compared with classic distillation.

3.8.2 *Retrofit to Cyclic Distillation*

Because cyclic distillation can improve the separation and energy efficiency and increase the capacity substantially, it can be considered as an attractive option in the retrofit of a conventional distillation sequence. Thus far, no paper has provided details of a retrofit using cyclic distillation, but it can be implemented easily in existing columns by simply changing the internals and the operating mode, thereby bringing new life to old towers by significantly increasing the column throughput, reducing the energy requirements, and offering better separation performance (Maleta et al., 2011; Patrut et al., 2014; Maleta et al., 2015). The detail studies of retrofit using cyclic distillation need be considered further, as well as how the column hydraulic changes, and how to overcome the limitation related to number of trays and the new column internal which, is more suitable for the operation of cyclic distillation.

3.9 Concluding Remarks

Improving the energy efficiency and/or increasing capacity of existing distillation has attracted increasing attention even though it is a well-defined separation process. This chapter discussed several attractive techniques that can be applied to a single column in retrofit or debottlenecking. The signals to apply each technology are described so that they can be chosen quickly and precisely in retrofit projects. Recent studies and industrial applications have shown that multi-effect distillation, heat pump-assisted distillation and self-heat recuperation techniques reduce significantly the amount of energy required for the separation of mixtures into pure product streams among those techniques. On the other hand, at this moment, the main barrier associated with heat pump applications is not theory but hardware and concerns, which will require further consideration.

References

Agrawal R (2000) Multieffect distillation for thermally coupled configurations. AIChE 46:2211–2224

Amminudin KA, Smith R (2001) Design and optimization of fully thermally coupled distillation columns. Part 2: Application of dividing wall columns in retrofit. Trans IChemE 79:716–724

Annakou O, Mizsey P (1995) Rigorous investigation of heat pump assisted distillation. Heat Recov Syst and CHP 15:241–247

Aspen Technology (2009) Aspen HYSYS thermodynamics COM interface. Version number V7.1

Bruinsma D, Spoelstra S (2010) Heat pumps in distillation. Distillation Absorption, 21–28.

Capps RW (1993) Consider the ultimate capacity of fractionation trays. Chem Eng Prog. 89:37–42

Castillo FJL, Dhole VR (1995) Pressure analysis of the ethylene cold-end process. Com Chem Eng 19:89–94

Cheng HC, Luyben WL (1985) Heat-integrated distillation columns for ternary separations. Ind Eng Chem Process Des Dev. 24:707–713

Chew JM, Reddy CCS, Rangaiah GP (2014) Improving energy efficiency of dividing-wall columns using heat pumps, organic rankine cycle and kalina cycle. Chem Eng Process 76:45–59

Dejanović I, Matijašević Lj, Olujić Ž (2010) Dividing wall column-a breakthrough towards sustainable distilling. Chem Eng Process 49:559–580

Douglas JM (1988) Conceptual design of chemical processes. McGraw-Hill, New York

Engelien HK, Skogestad S (2005) Multi-effect distillation applied to an industrial case study. Chem Eng Process. 44:819

Fidkowski ZT, Agrawal R (1995) Utilization of waste heat stream in distillation. Ind Eng Chem Res 34:1287–1293

Fonyo Z, Krrat R, Rippin DWT et al (1995) Comparative analysis of various heat pump schemes applied to C splitters. Com Chem Eng: 19:1–6

Gadalla MA (2009) Internal heat integrated distillation columns (iHIDiCs)-New systematic design method. Chem Eng Res Des. 87:1658–1666

Gadalla MA, Olujic Z, Jansens PJ, Jobson M, Smith R (2005) Reducing CO emissions and energy consumption of heat-integrated distillation systems. Environ Sci Technol 39:6860–6870

Gaska RA, Cannon MR (1961) Controlled cycling improves various processes. Ind Eng Chem. 53:629–629

Göktun S (1995) Selection of working fluids for high-temperature heat pump. Energy 20:623–625

Humphrey JL, Keller II GE (1997) Separation process technology. McGraw-Hill, New York

Jobson M (2014) Energy considerations in distillation. In: Gorak A, Sorensen E (eds) Distillation: fundamentals and principles, 1st edn. Elsevier, New York, pp 225–270

Kansha Y, Kotani Y, Aziz M et al (2013) Evaluation of a self-heat recuperative thermal process based on thermodynamic irreversibility and exergy. J Chem Eng Japan 46:87–91

Kansha Y, Tsuru N, Fushimi C et al (2010) Integrated process module for distillation processes based on self-heat recuperation technology. J Chem Eng Japan 43:502–507

Kansha Y, Tsuru N, Sato K et al (2009) Self-heat recuperation technology for energy saving in chemical processes. Ind Eng Chem Res. 48:7682–7686

Kiss AA (2013) Advanced distillation technologies: design, control and applications, 1st edn. Wiley, West Sussex

Kiss AA (2014) Distillation technology – still young and full of breakthrough opportunities. Chem Technol Biotechnol 4:479–498

Kiss AA, Landaeta SJF, Ferreira CAI (2012) Towards energy efficient distillation technologies – making the right choice. Energy 47:531–542

Kister HZ (1992) Distillation design. McGraw-Hill, New York

Klemola KT, Ilme JK (1996) Distillation efficiencies of an industrial-scale i-butane/n-butane fractionator. Ind Eng Chem Res 35:4579–4586

Liu ZY, Jobson M (1999) The effect of operating pressure on distillation column throughput. Com Chem Eng. 23:S831–S834

Liu ZY, Jobson M (2004) Retrofit design for increasing the processing capacity of distillation columns. 2. Proposing and evaluating design options. Chem Eng Res Des 82:10–17

Long NVD, Lee MY (2012) Improvement of the deethanizing and depropanizing fractionation steps in NGL recovery process using dividing wall column. J Chem Eng Japan 45: 285–294

Long NVD, Lee MY (2013a) Design and optimization of heat integrated dividing wall columns for improved debutanizing and deisobutanizing fractionation of NGL. Korean J Chem Eng 30:286–294

Long NVD, Lee MY (2013b) A novel NGL (natural gas liquid) recovery process based on self-heat recuperation. Energy 57:663–670

Long NVD, Lee MY (2013c) Debottlenecking the retrofitted thermally coupled distillation sequence. Ind Eng Chem Res 52:12635–12645

Long NVD, Lee MY (2014) Review of retrofitting distillation columns using thermally coupled distillation sequences and dividing wall columns to improve energy efficiency. J Chem Eng Japan 47:87–108

Long VDN, Lee MY (2017) Novel acid gas removal process based on self-heat recuperation technology. Int. J. Greenh. Gas Control 64:34–42

Luyben WL (2016) Distillation column pressure selection. Sep and Purif Technol. 168:62–67

Lynd LR, Grethlein HE (1986) Distillation with intermediate heat pumps and optimal sidestream return. AIChE J. 32:1347–1359

Maleta VN, Kiss AA, Taran VM et al (2011) Understanding process intensification in cyclic distillation systems. Chem Eng Process. 50:655–664

Maleta BV, Shevchenko A, Bedryk O et al (2015) Pilot-scale studies of process intensification by cyclic distillation. AIChE J. 61:2581–2591

Manley DB (1998) Close-coupling of interreboiling to recovered heat. U.S. Patent No. 5,755,115, 26 May

Matsuda K, Kawazuishi K, Kansha Y et al (2011) Advanced energy saving in distillation process with self-heat recuperation technology. Energy 36:4640–4645

Mizsey P, Fonyo Z (1992) Energy integrated distillation system design enhanced by heat pumping. Distillation Absorption, 1369–1376

Moser F, Schnitzer H (1985) Heat pumps in industry. Elsevier, Amsterdam

Olujić Ž, Sun L, de Rijke A, Jansens PJ (2006) Conceptual design of an internally heat integrated propylene-propane splitter. Energy 31:3083–3096

Patrut C, Bildea CS, Lita I et al (2014) Cyclic distillation—design, control and applications. Sep Purif Technol 125:326–336

Pleşu V, Ruiz AEB, Bonet J, Llorens J (2014) Simple equation for suitability of heat pump use in distillation. In: Klemeš J, Varbanov P, Liew PY (eds) Computer aided chemical engineering. Elsevier, Amsterdam, pp 1327–1332

Ranade S, Chao Y (1990) Industrial heat pumps: where and when? Hydrocarbon Process 71–73

Rivera-Ortega P, Picón-NúñezM, Torres-Reyes E et al (1997) Thermal integration of heat pumping systems in distillation columns. Appl Therm Eng. 19:819–829

Robinson CS, Gilliland RE (1950) Elements of fractional distillation, 4th edn. McGraw-Hill, New York, p 169–175

Soave G, Feliu JA (2002) Saving energy in distillation towers by feed splitting. Appl Therm Engi 22:889–896

Suphanit B (2011) Optimal heat distribution in the internally heat-integrateddistillation column (HIDiC). Energy 36:4171–4181

Turton R, Bailie RC, Whiting WB et al (2012) Analysis, synthesis and design of chemical processes, 4th edn. Prentice Hall, Upper Saddle River

Wankat PC, Kessler DP (1993) Two-feed distillation: same-composition feeds with different enthalpies. Ind. Eng. Chem. Res. 32:3061–3067

Zimmerman H, Walzl R (2009) Ethylene. In Ullmann's encyclopedia of industrial chemistry, 7th edn. Wiley, Weinheim

Chapter 4
Promising Retrofit Technologies for Multi-Column System

4.1 Introduction

To improve the performance for multi-column system, all promising retrofit technologies for single column can be employed. Besides, process intensification is currently considered the main trend for improving the process performance together with process integration. Process intensification aims to reduce the number of equipment units through the innovative design of multifunctional equipment that leads to a decrease in investment cost and significant energy savings (Rong and Turunen, 2006). For a distillation system, the thermal coupling technique provides a new mechanism for retrofitting the traditional simple column configuration through the principle of process intensification (Errico et al., 2007). This occurs by eliminating either the condenser or reboiler or both in the traditional simple column configuration by thermal coupling. Thermal coupling will provide simultaneous mass and heat transfer between the different tasks in the separation sequence. Therefore, thermal coupling technique a unique approach for intensifying the traditional simple column configuration, and has several advantages: (a) decrease in the number of the heat exchangers, (b) improved separation efficiency, (c) potential to save both energy and capital costs, (d) potential to increase the capacity, and (e) reduce CO_2 emissions. Especially, combination of heat integration and process intensification or combination of external and internal heat integration is expected to bring a substantial improvement.

4.2 Re-Arranging the Distillation Sequence

4.2.1 Introduction

For a ternary mixture, the direct and indirect conventional distillation sequence can be employed to separate a three component feed into pure products, which is illustrated in Fig. 4.1a and 4.1b, respectively. Every distillation column has a

N.V.D. Long, M. Lee, *Advances in Distillation Retrofit*, DOI 10.1007/978-981-10-5901-8_4

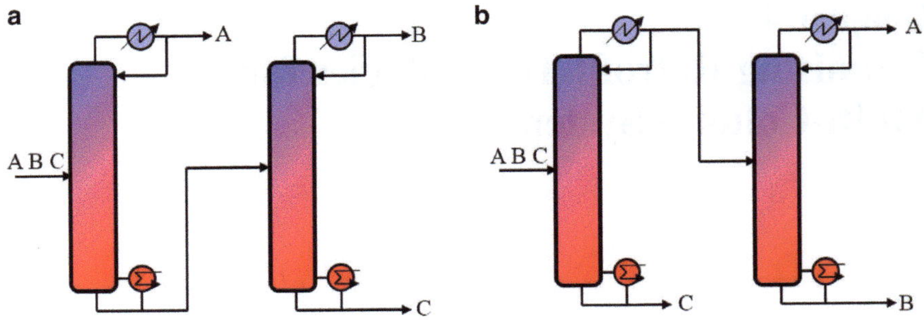

Fig. 4.1 Schematic diagram of the (a) direct sequence and (b) indirect sequence

reboiler and a condenser. The number of product flows in the columns is always the same, two. These two sequences are well-known, easy to control and operate, and familiar to operators. There may be a substantial difference in operating cost between the direct and indirect sequence. Before considering the heat integration and/or process intensification of distillation in a retrofit, a re-arrangement of the existing sequence needs to be considered. In particular, an existing indirect sequence can be retrofitted to a direct sequence and reversely or the existing sequence can be rearranged. The following useful heuristics (Smith, 2005) might be used to analyze and evaluate a retrofit:

Heuristic 1. Separations where the relative volatility of the key components is close to unity or that exhibit azeotropic behavior should be performed in the absence of nonkey components. In other words, perform the most difficult separations last.

Heuristic 2. Sequences that remove the lightest components alone one by one in column overheads should be favored. In other words, favor the direct sequence.

Heuristic 3. A component comprising a large fraction of the feed should be removed first.

Heuristic 4. Favor splits in which the molar flow between the top and bottom products in individual columns is as equal as possible.

Furthermore, the removal of corrosive components, reactive components, and monomers as soon as possible (Douglas, 1988) should be considered during retrofit analysis. This is because the material from the corrosion of a column is much more expensive than carbon steel if corrosive components are present. In addition, reactive components will change the separation problem, while monomers might foul reboilers, which need to operate the column under expensive vacuum conditions (Douglas, 1988).

Normally, a direct sequence consumes less energy compared to indirect sequence. When an indirect sequence is used, changing a total condenser to partial condenser in the first column should be considered to save energy. Just a few modifications are carried out compared to heat integration and/or process intensification.

Figure 4.2 shows the intermediate component composition profile in the direct and indirect sequences. Normally, in the direct sequence, the composition of B in the first column increases below the feed as the more volatile component A decreases. On the other hand, moving further down the column, the composition

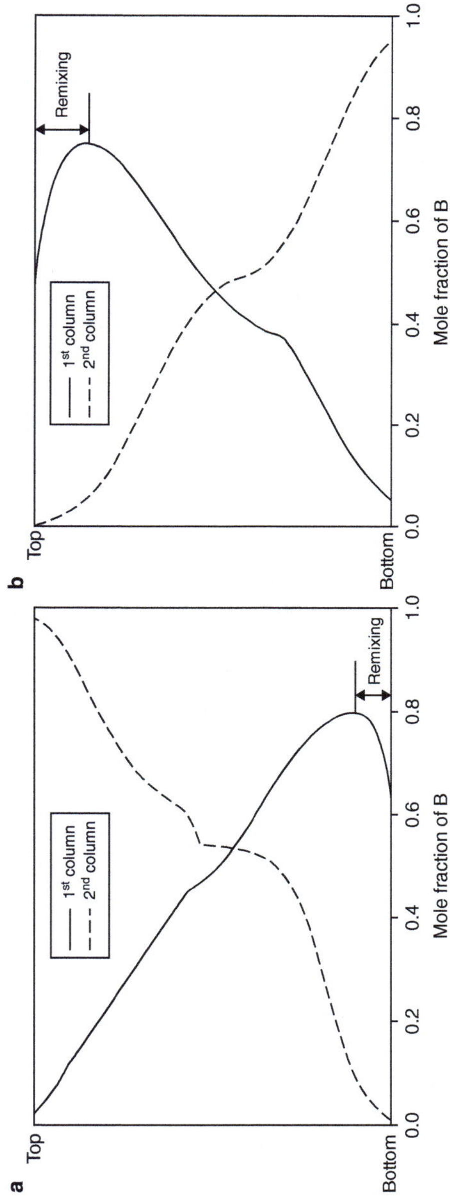

Fig. 4.2 Composition profile of B in the (a) direct sequence and (b) indirect sequence

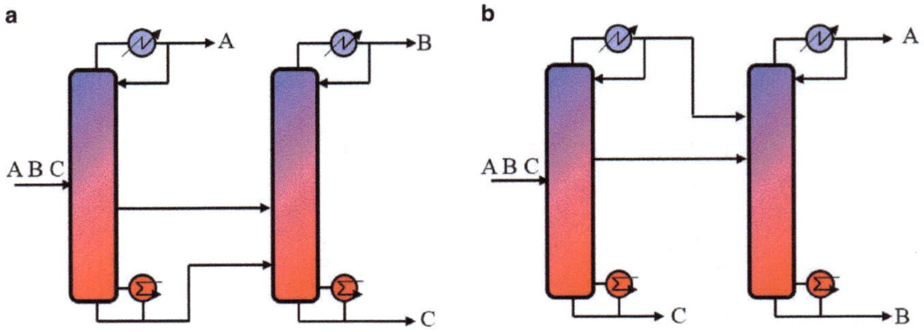

Fig. 4.3 Schematic diagram of the (a) modified direct sequence and (b) modified indirect sequence

of component B decreases again as the composition of the less-volatile component C increases (as shown in Fig. 4.2a). Therefore, the composition of B reaches a peak only to be remixed (Smith, 2005). Similarly, with the first column in the indirect sequence, the composition of component B first increases above the feed as the less-volatile component C decreases (Smith, 2005). This reaches a maximum only to decrease with an increase in the more volatile component A (as shown in Fig. 4.2b). Again, the composition of component B reaches a peak only to be remixed. The remixing that occurs in both sequences is a source of inefficiency in the separation.

In that case, side-stream columns (SSCs), in which the intermediate component is withdrawn by a side draw (Fig. 4.3), are considered to more energetically favorable alternatives to utilize this remixing phenomenon, possibly leading to substantial energy consumption in the two-column sequences. In particular, the first column can be retrofitted to a SSC, and then the side stream also is fed into the second column. Therefore, the SSCs are used to retrofit in a cost-effective manner, where a highly pure product is not required. Note that to optimize the retrofitted first column, the flow rate and location of the side stream are varied. The side stream is normally withdrawn at the tray where composition of intermediate component reaches the peak of the composition profile. Moreover, the location of that stream in the second column also need to be optimized so that its composition is matched to the feed tray composition in the second column to reduce the mixing effect.

4.2.2 Case Study

4.2.2.1 Existing Conventional Distillation Sequence

A debutanizer with 40 theoretical trays was designed for use at 5.0 bar (Long and Lee, 2015) (Fig. 4.4). A deisobutanizer, possessing 92 theoretical trays (Amminudin and Smith, 2001), was designed and operated at 6.5 bar (Fig. 4.4) because commercial isobutane can be condensed with cooling water under this pressure (Klemola and Ilme, 1996). The base case simulation model showed that the energy consumption by the debutanizer and deisobutanizer were 3165 and 6983 KW, respectively.

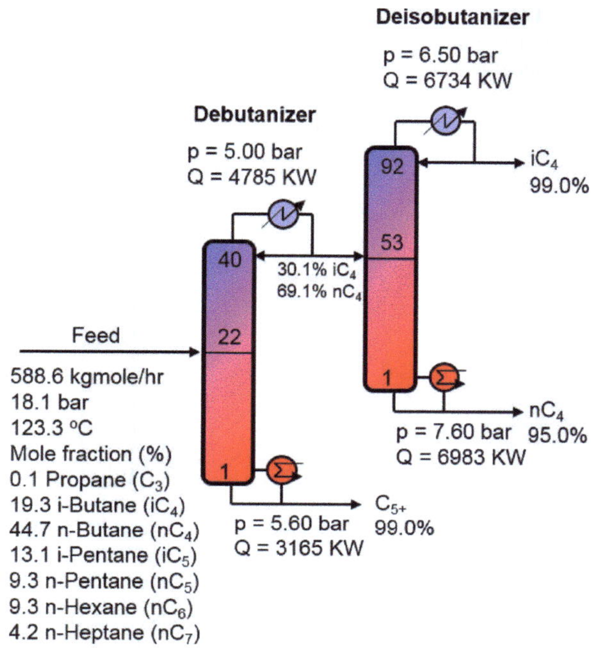

Fig. 4.4 Simplified flow sheet illustrating the existing separation train of two conventional columns

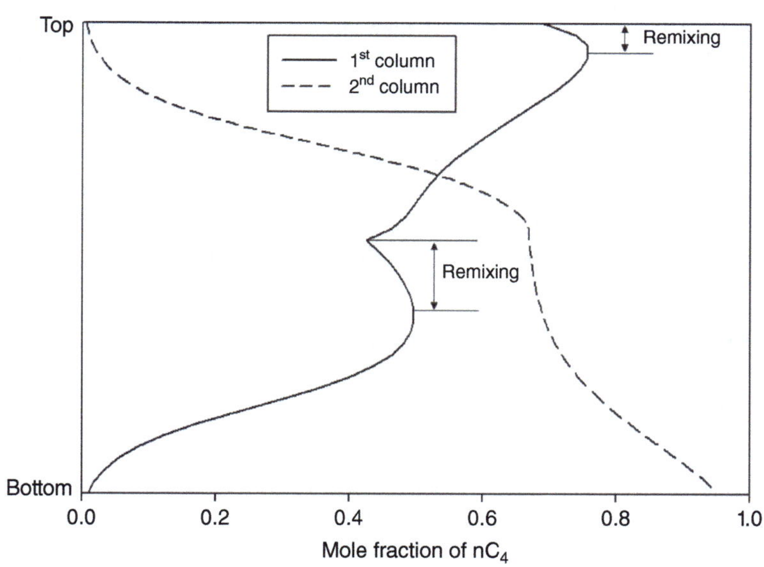

Fig. 4.5 Composition profile of nC_4 in the existing indirect sequence

4.2.2.2 Retrofitted Distillation Sequence

Figure 4.5 presents the composition profile of nC_4 in the existing indirect sequence. The composition of nC_4 in the debutanizer increases from the bottom as

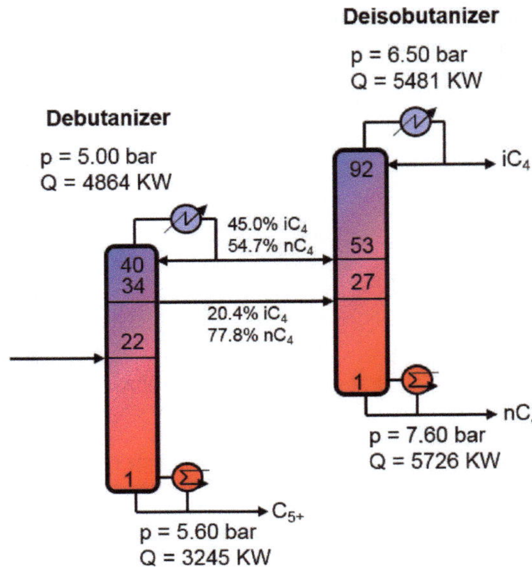

Fig. 4.6 Simplified flow sheet illustrating the retrofitted sequence

the less-volatile component C decreases. The profile has one remixing at several trays around the feed location because there is some mismatch between the feed composition and feed tray composition. The composition profile continues to increase until it reaches a peak and then decreases as the concentration of the more volatile component iC_4 increases. The remixing occurring in this sequence is a source of inefficiency in the separation, which can be overcome by retrofitting the debutanizer to a SSC (Fig. 4.6). In particular, a side liquid stream is withdrawn at the peak of nC_4, and then fed into the deisobutanizer to reduce its duty. The condenser duty and reboiler duty savings generated from this arrangement were 10.2% and 11.6%, respectively, which were comparable to the performance of the existing column arrangement. The modification is simple but effective.

4.3 Prefractionator Arrangement

4.3.1 Introduction

As discussed above, although the control and operation strategy for the conventional columns is simple, it is inefficient in terms of the energy due to the mixing entropy by irreversible split (Asprion and Kaibel, 2010).To improve the energy efficiency, a more novel distillation scheme called the prefractionator arrangement, was proposed, as illustrated in Fig. 4.7. In this sequence, there is one extra product flow in the second column of the column sequence that is known as a side stream (Malinen and Tanskanen, 2009). This prefractionator arrangement takes advantage of the control and operation, whereby control of the external flows from the reboiler and condenser of prefractionator column is facile.

The prefractionator configuration is advantageous to conventional distillation schemes and shows better energy performance (Amminudin and Smith, 2001; Poth et al., 2004).

4.3.2 *Case Study*

As shown in Fig. 4.8, nC_4 is split crudely in the prefractionator (debutanizer). This column is used to obtain iC_4 and nC_4 in the distillate stream, while nC_4 and C_{5+} are collected in the bottom stream. Note that the operating pressure of the debutanizer is increased so that the vapor stream from this column can enter the deisobutanizer without using a compressor. According to the rigorous simulation results, the use of a retrofitted prefractionator arrangement can save 20.6% of the reboiler duty.

Fig. 4.7 Schematic diagram of the prefractionator arrangement

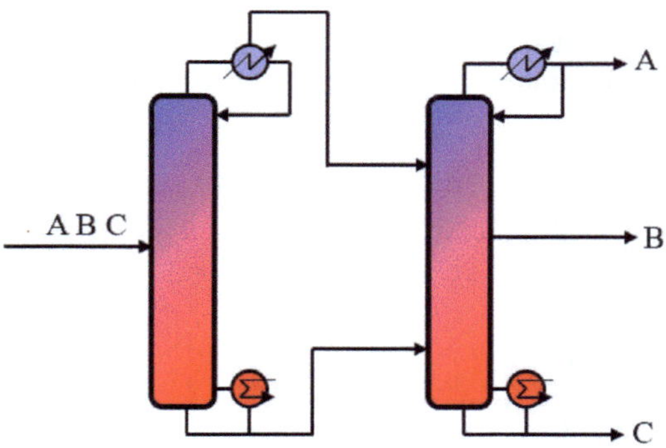

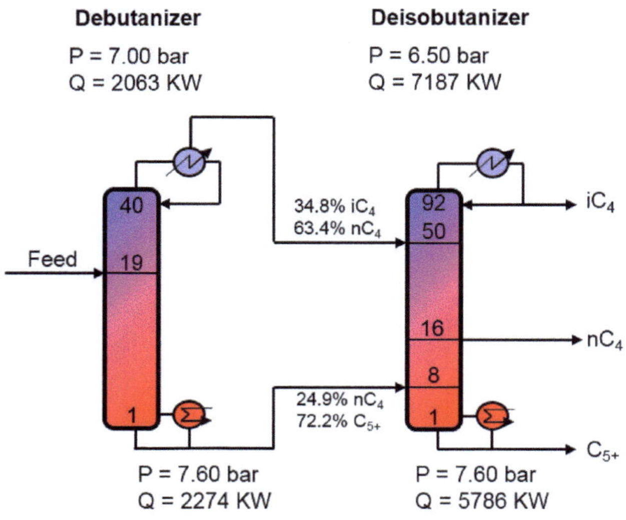

Fig. 4.8 Simplified flow sheet illustrating the retrofitted prefractionator arrangement

4.4 Thermally Coupled Distillation Sequence

4.4.1 Introduction

The thermally coupled distillation sequences (TCDSs) were obtained by the implementation of interconnecting streams (one in the vapor phase and the other in the liquid phase) between two columns (Long and Lee, 2012a). These resulted in an improvement of the energy efficiency by avoiding the remixing effect and the repeated separation path. Each interconnection also replaces one condenser or one reboiler from one of the columns, providing potential cost savings. Such systems have attracted considerable attention for the separation of ternary mixtures, and have been shown to provide significant energy savings with respect to both the conventional direct and indirect distillation sequences. The three most widely analyzed thermally-coupled systems are the sequence with a side rectifier, the sequence with a side stripper, and a fully thermally coupled system (or Petlyuk column), which are illustrated in Figs. 4.9a, b, and 4.10, respectively (Glinos et al., 1986; Triantafyllou and Smith, 1992; Hernández and Jiménez, 1996; Jiménez et al., 2003). These

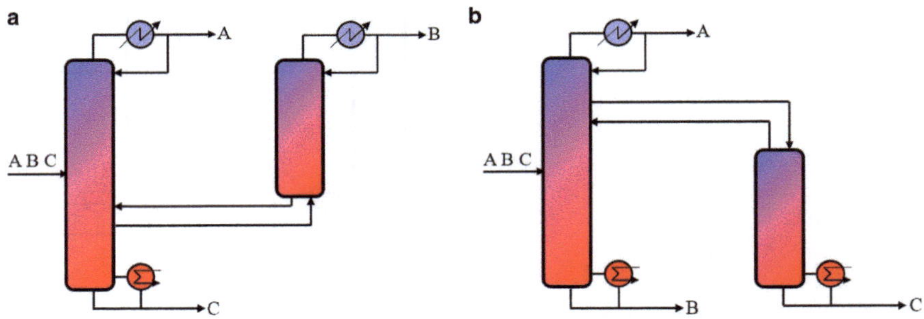

Fig. 4.9 Schematic diagram of the TCDSs: (a) system with a side rectifier and (b) system with a side stripper

Fig. 4.10 Schematic diagram of the Petlyuk column

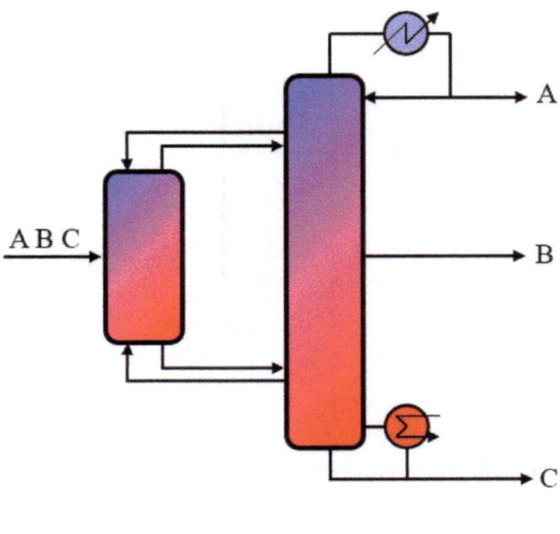

sequences have been shown to provide up to 30% energy savings compared to the conventional direct and indirect distillation sequences (Alatiqi and Luyben, 1985; Blancarte-Palacios et al., 2003). Most of the energy savings in the Petlyuk column arrangement are distributed to the prefractionator (Amminudin et al., 2001). A significant reduction in the remixing inefficiency is provided by a non-sharp split in the prefractionator (Triantafyllou and Smith, 1992). In addition, the prefractionator minimizes the mixing losses at the feed tray (Amminudin et al., 2001).

Although the energy savings from the Petlyuk column are well known, it undergoes strong interactions between the two columns because of their thermal integration, which creates difficulties in operation. In particular, the Petlyuk system includes a vapor stream that passes from the first column (or prefractionator) to the main column, and another that moves in the opposite direction. This situation requires that the pressure at the bottom of the prefractionator be lower than that of the main column, whereas the pressure at the top should be higher than that of the main column (Agrawal, 1999), which is believed to be difficult to control. Therefore, the TCDS with side rectifier and side stripper are mainly considered for a retrofit.

4.4.2 Retrofit to Thermally Coupled Distillation Sequence

4.4.2.1 Design

TCDS has attracted considerable attention in retrofit projects because of the lower energy requirements compared to existing conventional column sequences, as well as to the easy design and small modification (Long and Lee, 2013a). TCDS were obtained through the implementation of interconnecting streams (one in the vapor phase and the other in the liquid phase) between the two columns. Each interconnection replaces one condenser or one reboiler from one of the columns, providing potential cost savings. Figures 4.11 and 4.12 show the possibility of rearranging two existing columns into a TCDS with a side rectifier and side stripper, respectively. To highlight the optimal use of existing columns, no change in the diameter or the total number of stages of each column is assumed. Retrofitting the existing distillation sequence begins with the development of preliminary designs for complex systems and minimizes the heat duty supplied to reboilers through optimization procedures. A modification for a retrofit involves a change in the draw trays for the vapor and liquid stream. More pipe work will be needed for the thermal link and re-assignment of the eliminated reboiler (or condenser) to the main column.

On the other hand, retrofitting a conventional distillation sequence to a retrofitted TCDS can create a bottleneck in the column. Therefore, the hydraulics of retrofitted TCDS need to be checked. A new indicator of the hydraulic performance of an existing distillation column, the fractional utilization of area (FUA), is used. A more detailed description about the FUA can be found elsewhere (Liu and Jobson, 2004a). Figure 4.13 shows the FUA curve for an existing column sequence and the main column in the retrofitted TCDS, which indicates that flooding occurring in the main column. In particular, when retrofitting two column

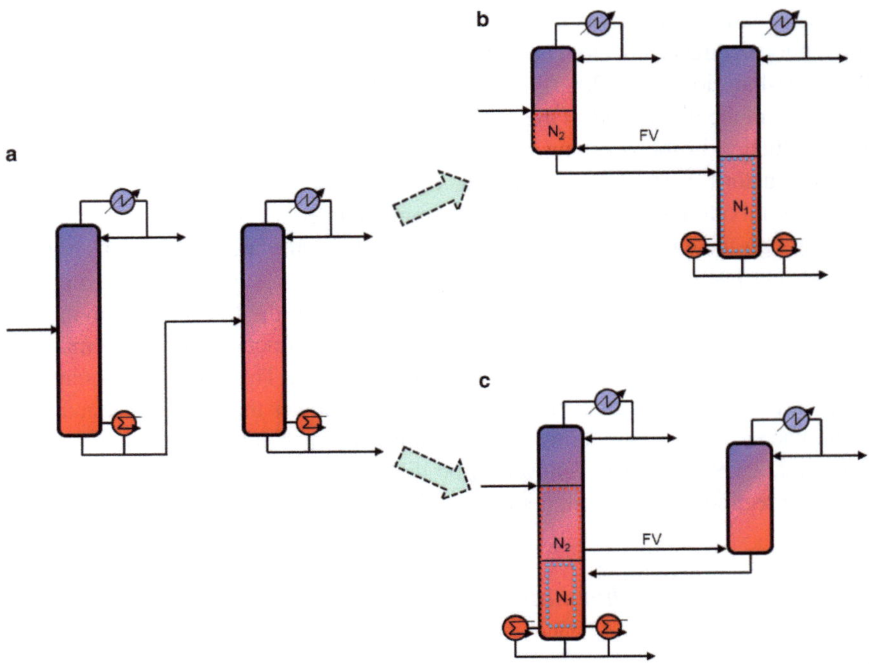

Fig. 4.11 Schematic diagram of the (a) existing direct sequence; (b) and (c) retrofitted TCDSs with a side rectifier

sequences to the TCDS with a side rectifier, the bottom section of the main column needs to cover all the vapor flows that are shared by the main column and side rectifier. This can create a bottleneck in the bottom section of the main column. As shown in Fig. 4.13a, the FUA value is larger than 1 in the stripping section of the main column in a retrofitted TCDS with side rectifier. Similarly, the top section of the main column needs to cover all the vapor flows, which are contributed by the main column and side stripper. These can cause bottleneck problems in the top section of the main column (Fig. 4.13b).

Frequently, the highest liquid and vapor traffic in a distillation tower is directly above the bottom reboiler (Reid, 2000). When the tower capacity is limited by these bottom section trays, the installation of a side reboiler sometimes allows an increase in capacity. A few process designs use a side reboiler to reduce the diameter of a new tower. A side reboiler will reduce the vapor and liquid traffic below the tray where the side reboiler is placed, and increase the vapor and liquid traffic above the tray (Liu and Jobson, 2004b), which will increase the area utilization above the side reboiler. Similarly, a side cooler will decrease the vapor and liquid traffic on the trays above it and increase the vapor and liquid traffic below it, which will increase the area utilization below the side condenser. A strategy using a side reboiler or a side condenser is proposed based on these phenomena. Figure 4.14 presents a schematic diagram of the TCDS with an added side reboiler or side condenser. In a retrofitted TCDS with side rectifier, a side reboiler is used to reduce the flows below the side reboiler, which will decrease and increase the area utilization below and above the side reboiler,

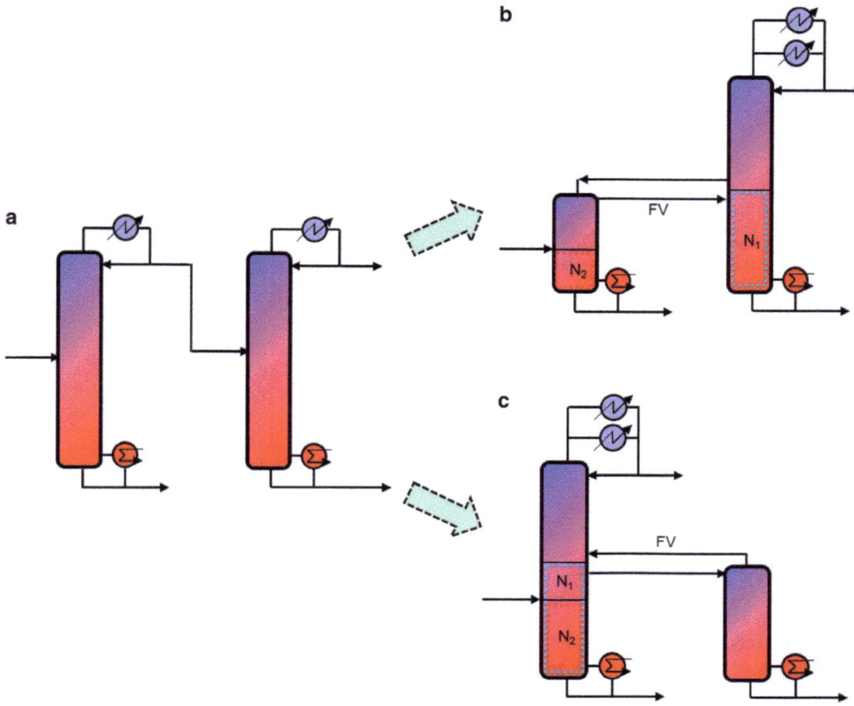

Fig. 4.12 Schematic diagram of the (a) existing indirect sequence; (b) and (c) retrofitted TCDSs with a side stripper

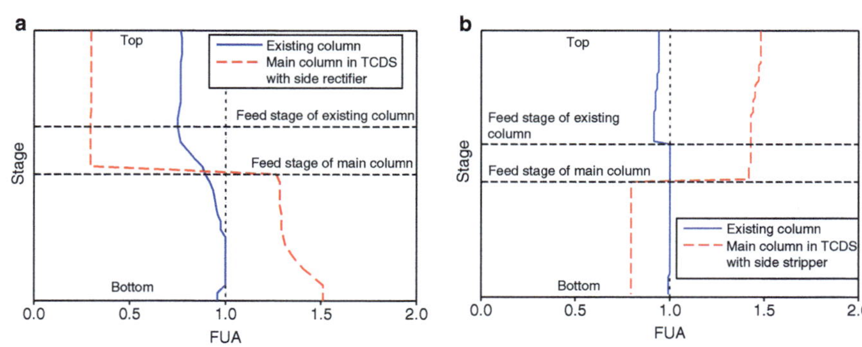

Fig. 4.13 FUA curve for (a) existing column in the direct sequence and main column in a retrofitted TCDS with side rectifier (b) existing column in the indirect sequence and main column in a retrofitted TCDS with side stripper

respectively. Similarly, a bottleneck often occurs in the rectifying section of the main column in a retrofitted TCDS with side stripper. A side condenser is used to reduce the area utilization above the side condenser.

A strategy for retrofitting a conventional distillation sequence to the TCDS is proposed. Figure 4.15 presents the steps of the retrofit strategy. The optimal retrofitted

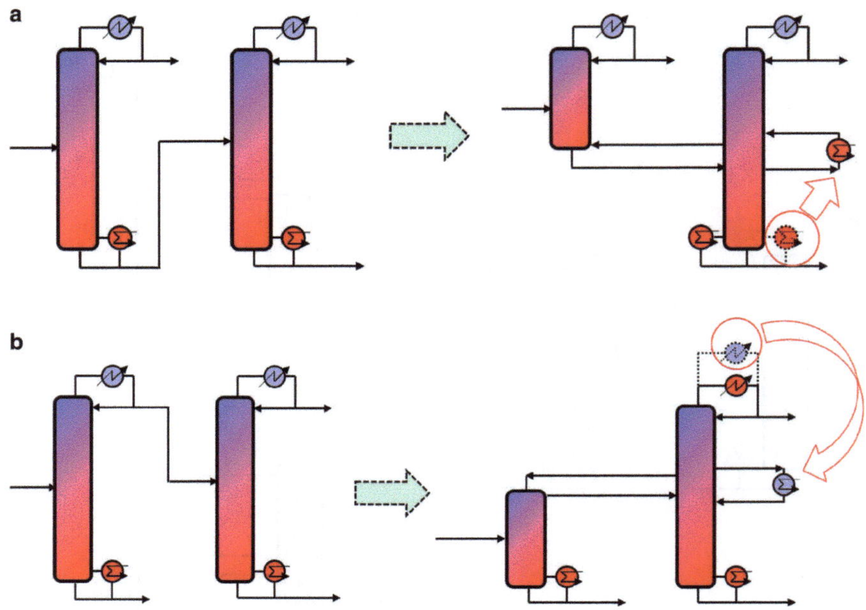

Fig. 4.14 Schematic diagram of the TCDSs: (a) system with a side rectifier and side reboiler; and (b) system with side stripper and side condenser

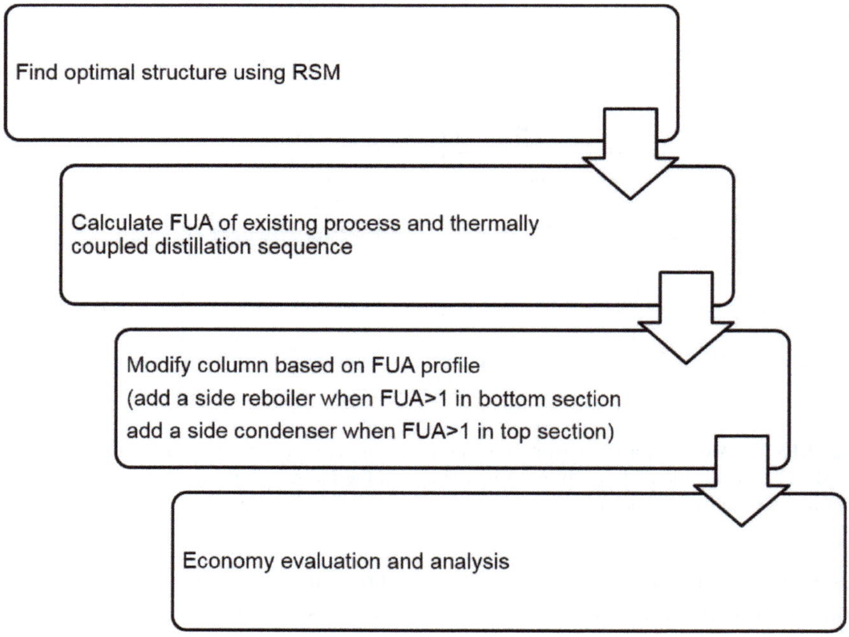

Fig. 4.15 Flow diagram of the retrofit design strategy to the TCDS

TCDS structure needs to be obtained. The FUA tool is then used to determine if the optimal retrofitted TCDS is flooded. Next, a side reboiler or a side condenser is used to remove the bottleneck in the main-column based on FUA analysis. A side reboiler can be used in cases where a bottleneck occurs at the bottom of a column. On the other hand, when a bottleneck occurs in the rectifying section, a side condenser can be installed to remove the bottleneck. Finally, the overall economics can be evaluated and analyzed to provide a final decision.

4.4.2.2 Optimization

The response surface methodology (RSM) is a general technique for an empirical study of the relationships between the measured responses and independent input variables (Box and Wilson, 1951). A response surface is normally a polynomial, whose coefficients are extracted by a simple least-square fit to the experimental and/or simulated data. The RSM is very powerful because, in addition to modeling, it can also optimize the conditions of the process under study (Box and Behnken, 1960; Mannan et al., 2007; Long and Lee, 2012b). To minimize the number of modifications in a retrofit, the number of trays of all columns was fixed to the same number in the conventional extractive distillation columns. After a preliminary design for the TCDS column, the main design variables, such as the vapor flow (FV) to the side rectifier, vapor (N1) and feed (N2) stream locations, were then optimized. The RSM was used to examine the effects on the reboiler. After determining the preliminary ranges of the variables through single-factor testing, a central composite design was used to determine how the variables interact and to optimize the system in terms of the reboiler duty.

The simulation data was fitted to a second-order polynomial model and the regression coefficients were obtained. The generalized second-order polynomial model used in response surface analysis can be expressed as follows:

$$Y = \beta_0 + \sum_{i=1}^{k} \beta_i X_i + \sum_{i=1}^{k} \beta_{ii} X_i^2 + \sum\sum_{i<j} \beta_{ij} X_i X_j + \varepsilon \tag{4.1}$$

where Y is the predicted response (reboiler duty), X_i are the uncoded or coded values of the variables, β_0 is a constant, β_i, β_{ii} and β_{ij} are the coefficients of the linear, quadratic and interactive terms, respectively, and ε is the error term. MINITAB software was used for response surface fitting and to optimize the reboiler duty.

4.4.3 Case Study

4.4.3.1 Existing Plant Subject to Retrofit Study

Figure 4.16 shows the conventional distillation sequence and its current operating conditions. With 18 theoretical stages, the deethanizer column produces a partially

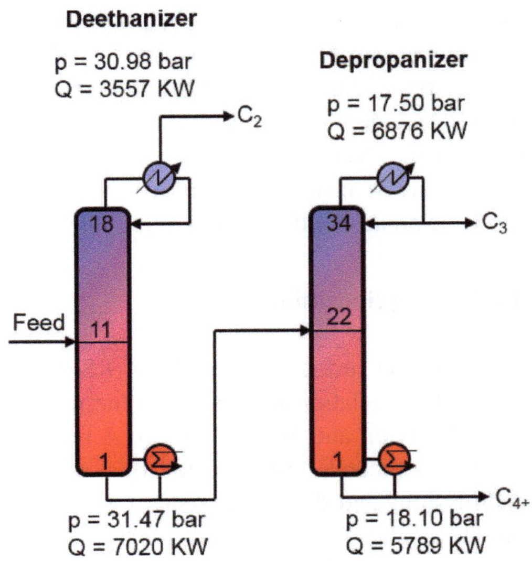

Fig. 4.16 Simplified flow sheet illustrating the existing separation train of two conventional columns

condensed overhead vapor stream. On the other hand, the depropanizer, which possesses 34 theoretical stages, was designed and operated at 17.5 bar because commercial propane (C_3) can be condensed with cooling water under this pressure (Manley, 1997). The Peng-Robinson equation of state, which supports the widest range of operating conditions and the greatest variety of the systems, was used to predict the vapor-liquid equilibrium of these simulations (Aspen Technology, 2009). The base case simulation model showed that the energy consumption of the deethanizer and depropanizer was 7020 and 5789 KW, respectively.

4.4.3.2 Retrofitted Thermally Coupled Distillation Sequence

To retrofit the conventional distillation sequences to the TCDS, the main design parameters of the existing columns were checked. The existing reboilers and condensers should also be checked for reusability with minimal modifications. Thirteen simulations were run to optimize two parameters of the TCDS structure. Figure 4.17a presents the three-dimensional response surface plot of the interaction between the main design variables, N1 and N2, which were plotted on the X and Y axes, respectively. The reboiler duty was plotted on the Z axis. Figure 4.17b shows the contour plot. The resulting second-order polynomial model is as follows:

$$Y = 9119.17 - 69X_1 - 9X_2 + 113.9X_1^2 + 698.9X_2^2 - 105.75X_1X_2 \quad (4.2)$$

where Y is the predicted response (reboiler duty), X_1 and X_2 are the coded values of the vapor and feed stream locations.

The quality of the polynomial model is expressed by the coefficient of determination, namely, R^2. Based on the analysis results from MINITAB, the model reports a

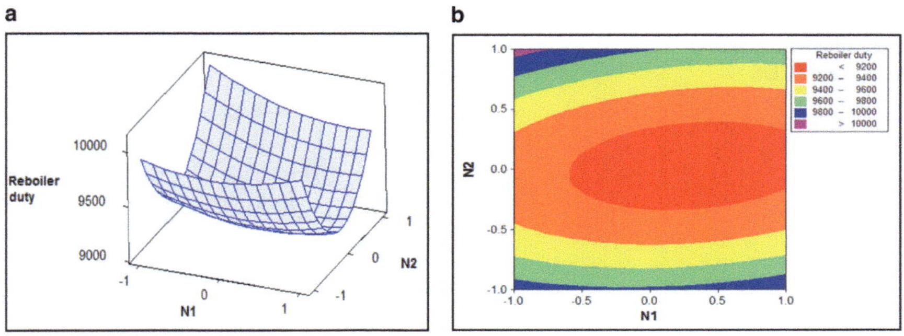

Fig. 4.17 (a) Three-dimensional response surface plot and (b) contour plot of interaction between N1 and N2

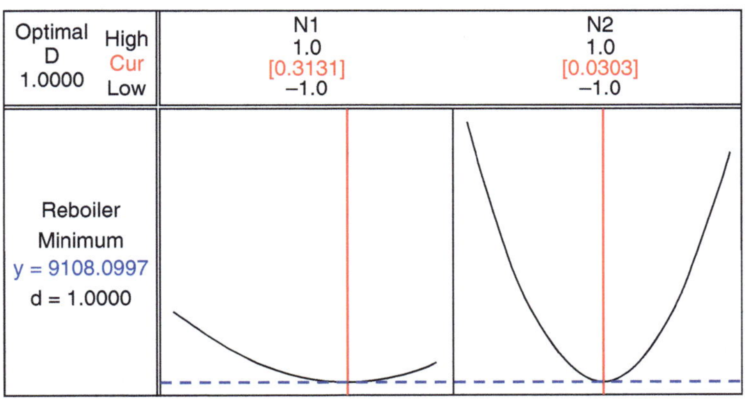

Fig. 4.18 Optimization plot by the RSM

high R^2 value of 0.988. The value of R^2 is close to 1.0, which is very high and advocates a strong correlation between the observed and predicted values. This indicates that the regression model provides an excellent explanation of the relationship between the independent variables and the response. The smallest reboiler duty was observed at the coded levels of the number of trays in the vapor stream and feed stream locations of 0.3131 and 0.0303, respectively (Fig. 4.18).

Under these conditions, the minimum reboiler duty by the retrofitted TCDS was predicted to be 9108 KW. The natural values of the variables can be derived from the coded levels. The vapor flows to the side rectifier were then optimized to minimize the reboiler duty, whilst maintaining sufficient product purity. Figure 4.19 shows a simplified flow sheet illustrating the retrofitted TCDS. The simulation also showed that the TCDS consumes 9216 KW, which is in good agreement with the results predicted by the RSM. Considerable savings (28.1%) in reboiler energy can be achieved. A decrease in column pressure reduces the energy cost for the reboiler but the condenser needs to be operated at lower temperatures. In this study, the temperature of the top of the deethanizer was reduced from 13.38°C to −10.77°C. To consider the effects of

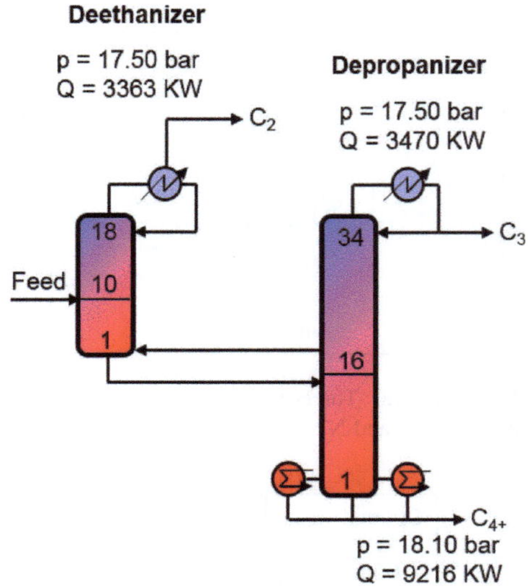

Fig. 4.19 Simplified flow sheet illustrating the retrofitted TCDS

different cooling temperatures on the overall economics, a different cost of refrigeration was used for the two cases (Table 3.1) (Turton et al., 2012). The operating cost was then calculated based on the sum of the costs of steam, refrigeration, electricity, and cooling water. The results showed that the operating cost can save up to 20.3% when retrofitting to TCDS. The reboiler in the deethanizer column can be used together with the existing reboiler in the depropanizer column.

On the other hand, when two columns are used to construct the TCDS, the existing depropanizer is likely to be bottlenecked, as shown in Fig. 4.20. In the FUA curve of the existing depropanizer, the fractional utilization of the area can vary from stage to stage, with some stages requiring all the available area (i.e., FUA = 1), whereas the others could accommodate significantly higher flows. The FUA value began to increase from the feed stage. In the main column of TCDS, the FUA increased dramatically from the vapor stream location (16th stage). The retrofitted depropanizer has a bottleneck problem in the stripping section. The FUA curve indicated an opportunity to reduce the area utilization in the stripping section of the column by increasing the flows in rectifying the section of the retrofitted depropanizer column.

4.4.3.3 Retrofitted Thermally Coupled Distillation Sequence with Side Reboiler

Side reboilers can be used to change the vapor and liquid flows within a column section to use the area available more uniformly (Liu and Jobson, 2004b). This will reduce the vapor and liquid traffic below the tray where the side reboiler is placed, and increase the vapor and liquid traffic above the tray. Therefore, the liquid and vapor traffic in the bottom-most section of the column can be reduced using a side reboiler, that is, the bottleneck problem can be removed.

Figure 4.21 shows the simplified flow sheet outlining the retrofitted TCDS with a side reboiler. As a result of using the side reboiler, the maximum area utilization

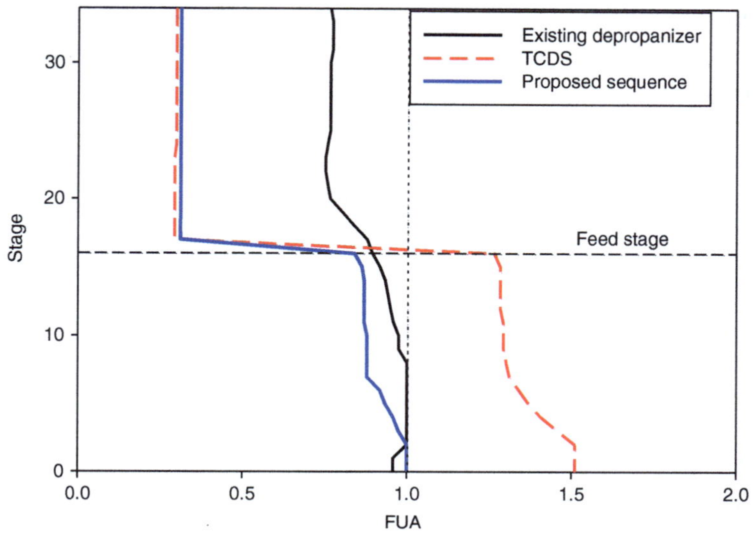

Fig. 4.20 FUA profiles of existing depropanizer, main column in TCDS, and main column in the proposed sequence

Fig. 4.21 Simplified flow sheet illustrating the retrofitted TCDS with a side reboiler

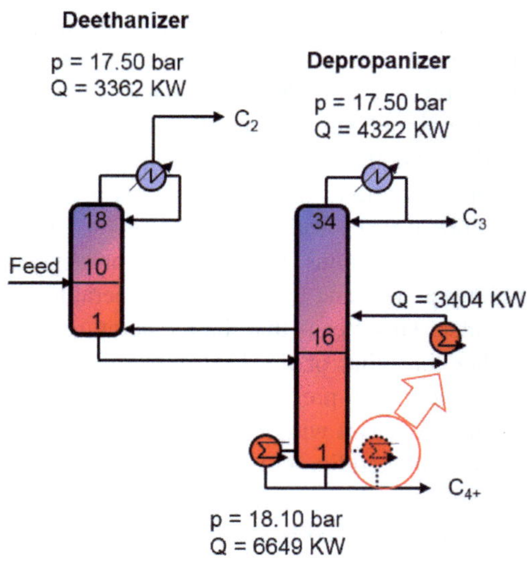

(in the stripping section) decreases to one and can be even smaller. When the maximum area utilization is less than one, the feed flow to the column can be increased until the maximum area utilization in the stripping section is one. This corresponds to an increase in throughput. The heat duty in the reboiler decreases with increasing heat duty in the side reboiler. Note that the reboiler of an existing deethanizer can be used as a reboiler in a retrofitted depropanizer, whereas the reboiler of the existing depropanizer is used as the side reboiler. As shown in

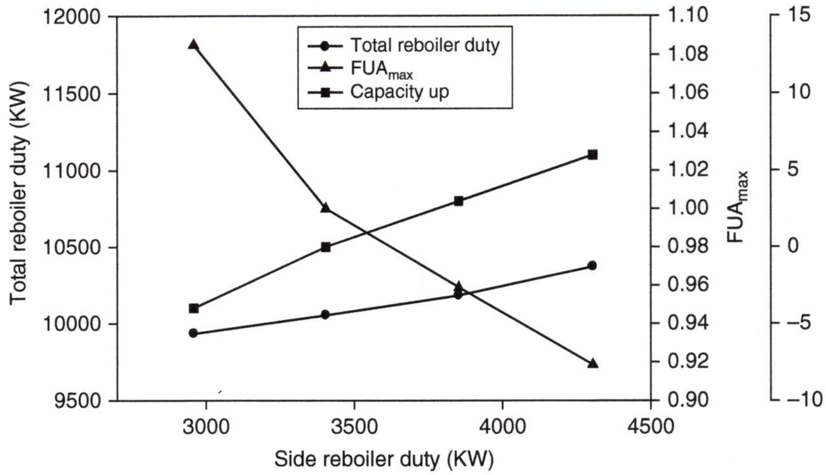

Fig. 4.22 Influence of side reboiler duty on the total reboiler duty, FUA_{max}, and capacity

Fig. 4.22, the FUA_{max} decreased with increasing heat duty in the side reboiler. On the other hand, the total duty increased with increasing duty in the side reboiler. The FUA_{max} and operating cost are a trade-off to be considered for this retrofit case. In this study, the aim was to decrease the energy requirements. Therefore, the chosen side reboiler duty was 3404 KW, which gives 21.5% savings in terms of the reboiler duty. More condenser duty is needed because the side reboiler boils up the liquid stream, which results in a heavier component (C_{4+}) in the top section. Another advantage of using a side reboiler is that the side reboiler in the stripping section can use lower-pressure steam because of the lower column temperature at that location (Luyben, 2004). This is particularly important when using distillation columns, in which the components being separated have very different boiling points, and hence large temperature differences between the condenser and reboiler. This means that the temperature of the heat source in the reboiler is much higher than that of the heat-removal sink in the condenser. A high temperature means that high-pressure steam is needed, or a fired reboiler, or a hot oil system is required in cases with temperatures above approximately 260 °C.

Note that adding either a side condenser or side reboiler moves one or more of the operating lines closer to the equilibrium curve in the classical McCabe-Thiele diagram (Reid, 2000). As a result, a revamp that reduces the tray traffic in a tower zone might normally require a larger number of equilibrium stages or operation with a higher total heat input and removal. On the other hand, using a side reboiler shifts some of the separation energy to a lower cost utility temperature level and also reduces the temperature difference for heat pump applications, which can reduce the overall energy costs.

Interestingly, an increase in the vapor flow rate to the side rectifier causes an increase in the composition of ethane (C_2). In this case, the composition of the contaminated heavy component in the propane product increases because the propane

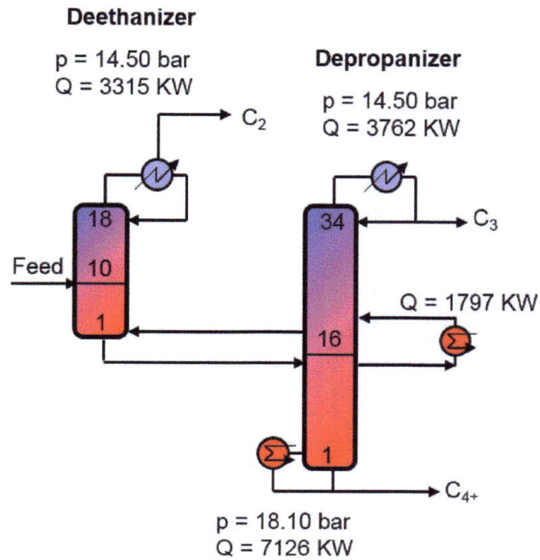

Fig. 4.23 Simplified flow sheet illustrating the retrofitted TCDS operated lower pressure with side reboiler

purity of the top product in the main column is kept constant. This causes an increase in the top temperature of the depropanizer, which creates a change of reduction the operating pressure. The pressure is possibly reduced to 14.5 bar instead of 17.5 bar. Figure 4.23 shows a simplified flow sheet illustrating the retrofitted TCDS operated at a lower pressure with a side reboiler. Compared to the conventional column sequence, this sequence can reduce the energy consumption by up to 30.3%.

The CO_2 emission reduction associated with the lower energy requirement is another important benefit of using the proposed sequence in the retrofit of a conventional distillation process. The method reported by (Gadalla et al., 2005) was used to estimate the CO_2 emissions. The study results showed that CO_2 emissions can be reduced dramatically when distillations are intensified by a thermal coupling technique and a part of the process heat is recovered for use instead of the primary steam. In particular, reductions of CO_2 emissions of up to 30.7% can be achieved.

4.5 Alternative Coupled Schemes of the Petlyuk Column

4.5.1 Introduction

Agrawal and Fidkowski (1998, 1999) proposed six alternate coupled schemes (Fig. 4.24) that might provide better operating properties than the Petlyuk column. In particular, the Petlyuk system shows a vapor stream that passes from the first column (or prefractionator) to the main column, and another that moves in the opposite direction. This situation requires that the pressure at the bottom of the prefractionator be lower than that of the main column, whereas the pressure at the

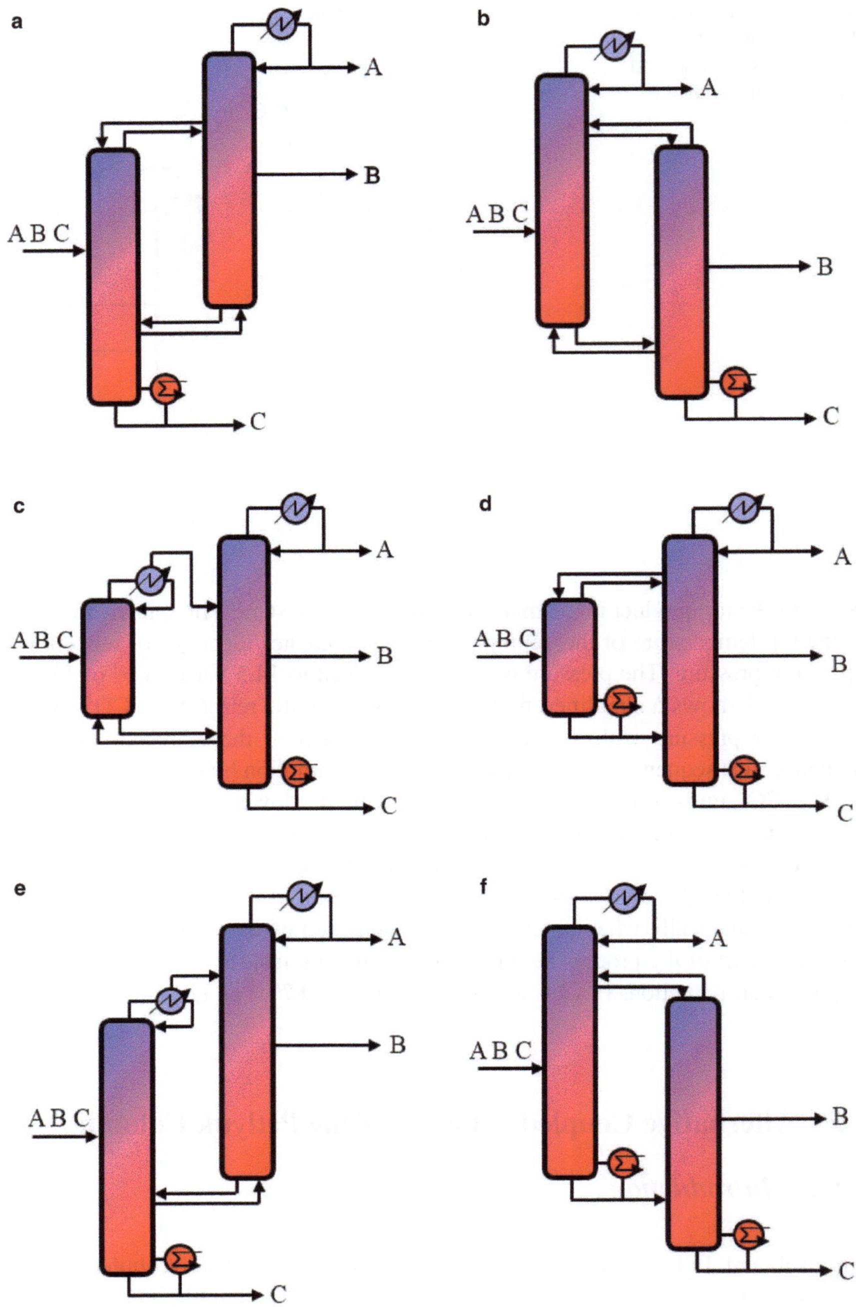

Fig. 4.24 Schematic diagram of the alternative coupled schemes of the Petlyuk column

top should be higher than that of the main column (Agrawal, 1999), which is perceived to be difficult to control. The six alternative distillation schemes have interesting potential in terms of their energy efficiency (Jiménez et al., 2003; Hernández et al., 2005). In the retrofit project, although savings should be considered, the operability is an important factor (Long and Lee, 2014). These alternative distillation schemes, which do not have the classical problem of controlling vapor transfer from one column to another, can be used in a retrofit.

4.6 Dividing Wall Column

4.6.1 Introduction

4.6.1.1 Three Products DWCs

One practical implementation of the Petlyuk column is the dividing wall column (DWC) that splits the middle section of a single vessel into two sections through the insertion of vertical wall in the vessel at the appropriate position – basically integrating the Petlyuk system into a single column shell (Kiss and Bildea, 2011). That vertical wall divides the column into a prefractionator and main section, as shown in Fig. 4.25. This arrangement is conceptually the same as the Petlyuk column given the thermodynamically equivalent arrangements (Amminudin et al., 2001). The top dividing wall column (TDWC) or bottom dividing wall column (BDWC) can be constituted moving the dividing wall to the top or bottom of the column, respectively. The DWC can effectively reduce the energy requirements by up to 30%, while also lowering the engineering and hardware costs compared to the conventional direct and indirect distillation sequences (Diggelen et al., 2010; Errico et al., 2009; Hernández & Jiménez, 1999; Kiss & Bildea, 2011; Long et al., 2010; Triantafyllou & Smith, 1992). Compared to the conventional two-column sequence, the DWC also can save capital expenditure and space with its single shell feature, single reboiler, and condenser. Furthermore, the space can be reduced, which is an important factor to consider in a retrofit design or some plant with space limitations.

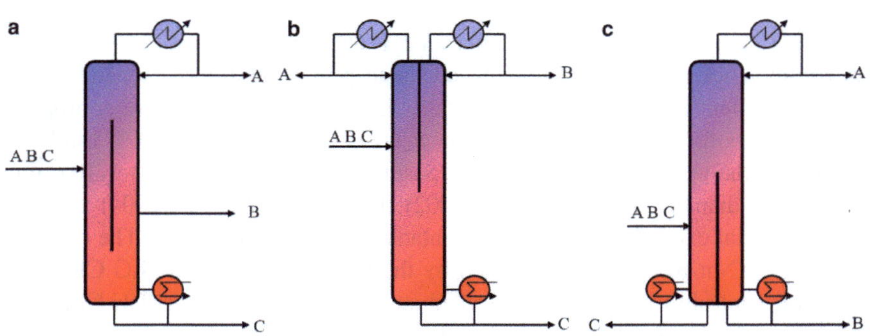

Fig. 4.25 Schematic diagram of the (a) DWC, (b) TDWC and (c) BDWC

For temperature-sensitive products, the thermal stress is reduced because the product is reboiled only once (Asprion and Kaibel, 2010). DWCs can also be used to conduct azeotropic (Kiss and Ignat 2012; Long and Lee, 2014), extractive (Long and Lee, 2011; Tututi-Avila et al., 2014) and reactive distillations (Kiss and Suszwalak, 2012) without any major changes to the column internals.

4.6.1.2 Limitations

On the other hand, there are many reasons why companies hesitate to apply DWCs. The design of a DWC adds additional complexity (Slade et al., 2006). The DWC design is more complex than conventional arrangements because of the larger number of design variables that interact with each other and need to be optimized simultaneously, making optimization both difficult and time consuming (Amminudin et al., 2001). Furthermore, it also has some application limitations, such as the only one operating pressure, a large temperature difference between the reboiler and condenser, heat transfer across the dividing wall, and the condition of high non-ideal behavior of the mixture, in which conventional or other configurations could be better suited. Large temperature in bottom can cause decomposition or polymerization of components (Long and Lee, 2014). In addition, the DWC may result in an increased column height compared to a configuration with several columns (Asprion and Kaibel, 2010). A further restraint is the perceived complexity and uncertainty of the control system. During DWC operation, the operator can only simply adjust the liquid split ratio by using an active liquid splitter. In most existing DWCs, vapor splitting is fixed in the design stage by a specific fixed open area for vapor flow and thus cannot be adjusted during column operation. In such a configuration, the vapor split ratio tends to respond inversely to changes in the liquid split ratio because of the variation in hydrostatic pressure of the liquid in each section. This tendency can cause an unfavorable loss in separation efficiency and result in higher reboiler duty to compensate for the loss in efficiency. Nevertheless, this problem is fixed recently by a hydraulic driven active vapor distributor, which is discussed in Sect. 4.6.4.

4.6.1.3 Industrial Applications

Interestingly, more than 10 years after first DWC application at BASF in 1985, other companies realized that making adequate provisions for the separation of a three component feed into pure products within a single shell, where appropriate, would save both operating and investment costs (Olujić et al., 2009). Linde AG constructed the world's largest DWC for Sasol, which was approximately107 m tall and 5 m in diameter (Schultz at al., 2002). Kolbe and Wenzel (2004) presented some industrial examples of the implementation of a DWC by Uhde. The first commercial DWC in Korea was developed by the PSDC team and the LG CRD team through the retrofit of a SSC and installed successfully in a plant at LG Chem Ltd. (Lee et al., 2011). Start-up of the commercial DWC was completed successfully. Changing the column configuration to a DWC resulted in improved product purity and significant energy savings (up to 40%) (Long and Lee, 2014). Significant advances have been made in this area with the successful exploitation of a DWC in retrofitting as pointed out by Slade et al. (2006). The DWC was implemented successfully and on schedule by Koch-Glitsch in 30 days during May 2005. Currently, there are more than 100 DWCs in operation, which are mainly new designs, but revamping existing columns will increase with time (Dejanović et al., 2010). Nowadays, many companies positively realize, accept, and study application of DWCs in their processes. More applications are expected, particularly as the value of energy conservation projects increase. According to Schultz et al. (2002), the DWC will become a standard distillation piece of equipment over the next 50 years.

4.6.1.4 Intensified Distillation Sequence including DWC for Quaternary Mixtures

Industrial separation problems often involve four or more components. Some work on extensions has been conducted toward the design of coupled systems for mixtures of more than three components (Vazquez-Castillo et al., 2009; Rong, 2011). Figure 4.26 shows a schematic diagram of five typical intensified distillation configurations for quaternary separations. They have the potential to save both energy and capital costs compared to conventional distillation sequence.

4.6.1.5 Four Products DWCs

Furthermore, the energy efficiencies of different Petlyuk arrangements, including the Kaibel, Sargent, and Agrawal arrangements have been confirmed (shown in

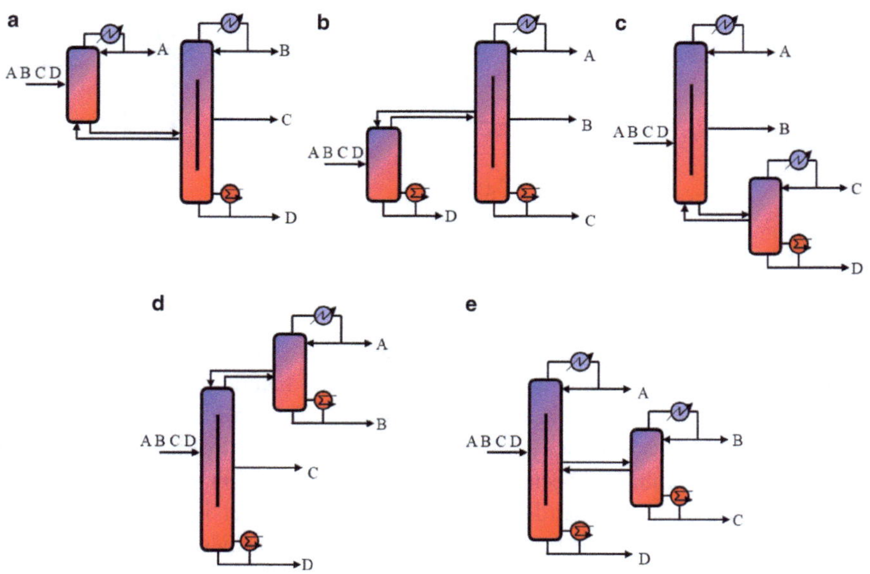

Fig. 4.26 Schematic diagram of five typical intensified distillation configurations for quaternary separations

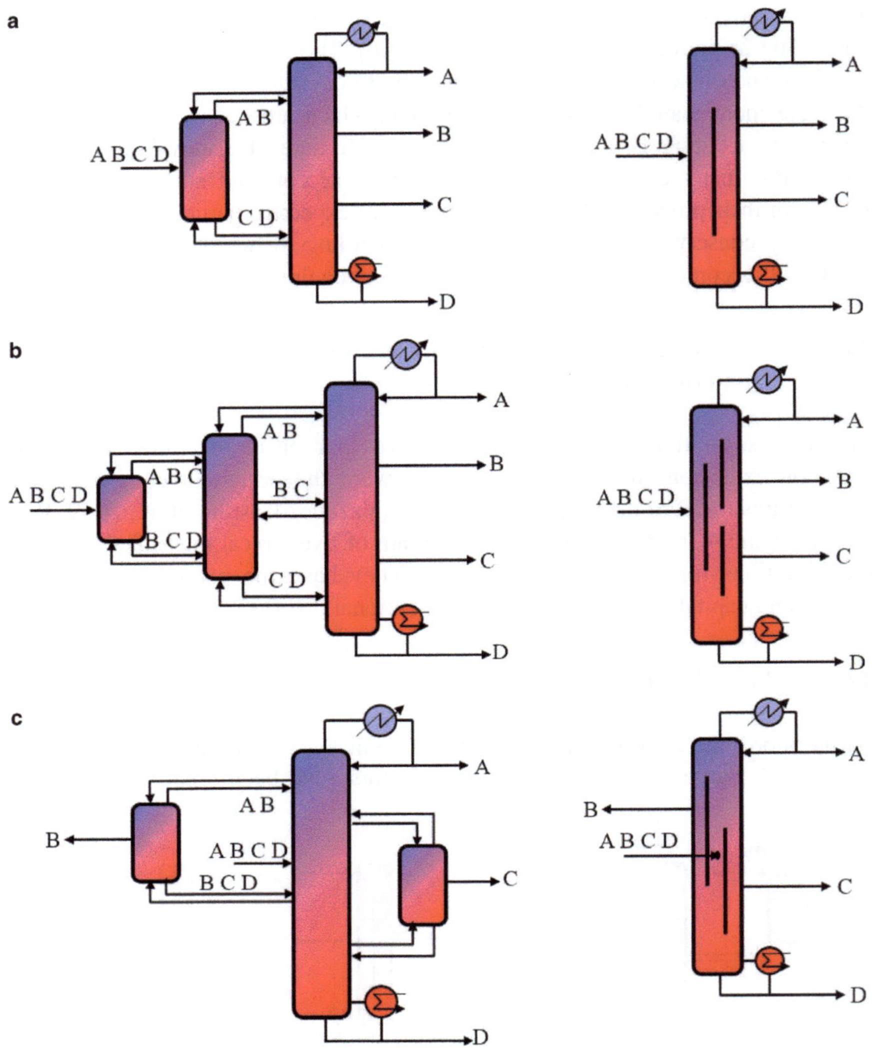

Fig. 4.27 Schematic diagram of the (a) Kaibel, (b) Sargent and (c) Agrawal arrangement

Fig. 4.27) (Christiansen et al., 1997). The Agrawal arrangement, which shows better results in this paper, was applied to improve the natural gas recovery process (Long and Lee, 2012c). The Kaibel arrangement is an interesting alternative taking into consideration its simplicity by adding a second side stream to a DWC. The design of this column has been mostly solved, but operation and control remains largely an open issue (Ghadrdan et al., 2011). The Kaibel is difficult to handle when minimizing the vapor flow rate at given product purities, while it appears to be easier to operate when the product purities are free. This can be used in refinery plants where the product purities do not play a vital role or when there is a bottleneck in the process. Recently, the Kaibel arrangement was further studied via an examination of the effects of vapor split manipulation (Ghadrdan et al., 2013).

4.6.2 Design and Optimization of DWC

Shortcut methods The design and optimization of a fully thermally coupled distillation column (FTCDC) are solved by a shortcut method, which uses a three-column model (Triantafyllou and Smith, 1992). The number of stages, minimum reflux, stage requirement, and feed stage are established using the conventional Fenske-Underwood-Gilliland-Kirkbride (FUGK). The best reflux ratio is then determined for a given cost scenario. The methods are restricted in their applications by simplifying the assumptions, such as constant molar flow and constant relative volatility. A new design procedure for the FTCDC was proposed to overcome the limitations of the previous approach (Amminudin et al., 2001). The new procedure adopted the equilibrium stage composition concept to define a liquid composition profile by solving the mass balance rigorously on each equilibrium stage, while restricting the procedure to the assumption of a constant molar flow. An assumption of constant relative volatility is not required because of its semi-rigorous nature.

V_{min} diagram method Halvorsen and Skogestad (2003) proposed the use of a minimum vapor flow, or so called V_{min} diagram method, which is a graphical representation of the minimum energy. In this method, a constant pressure, constant relative volatility, and constant molar flow under a limiting case with an infinite number of stages were assumed. Fidkowski and Krolikowski (1986) and Carlberg and Westerberg (1989a, b) presented the minimum-energy expressions based on the Underwood equations for the Petlyuk arrangements with three components, and Halvorsen and Skogestad (2003) added the minimum energy requirement for the general multicomponent case. An important finding was that the minimum energy requirement ($V_{min}^{Petlyuk}$) and the detailed vapor flow requirements may be found based on the V_{min} diagram.

Response surface method A practical method based on the shortcut and RSM was developed to design and optimize the structure of DWCs (Long and Lee, 2012b). A shortcut design based a conventional column configuration was used to initialize the DWC structure (Lee et al., 2016), which was then optimized efficiently using the RSM. Because of the large number of mixed type of variables in the design of a DWC, a statistical approach was found to be a powerful tool for determining the effects of the factors and their interactions (Long and Lee, 2012b).

Other rigorous methods Several rigorous methods for design and optimization of DWCs have been proposed, such as a rigorous design method based on the models of the detailed column superstructures (Dunnebier and Pantelides, 1999), a rigorous optimization method, where Aspen Plus™ software is linked to an external optimization routine based on an evolutionary algorithm (Wenzel and Rohm, 2003), a sequential quadratic programming-based method implemented in Aspen Plus™ (Kiss and Suszwalak, 2012), and a MILP approach based on the use of two superstructure levels (Caballero and Grossmann, 2014). Recently, Long et al. (2016) proposed a optimization methodology, in which Hysys is linked to Excel, which was found to be simple and effective for a complex system. This methodology can handle both structure and operating variables simultaneously.

4.6.3 *Retrofit Using DWC*

The aim of most retrofit projects for distillation columns is to reduce the energy requirements and/or to increase the capacity while maintaining the purity and recovery of the main product. The key to a successful retrofit lies in exploiting the existing hardware by maximizing the use of existing equipment while, at the same time, minimizing the need for new hardware to minimize the capital cost (Amminudin and Smith, 2001). Because DWCs represent a very promising technology, allowing a significant energy reduction, they can be considered to retrofit the conventional distillation sequence. Figures 4.28 and 4.29 show a schematic diagram of promising configurations for the retrofit using DWC.

Note that in cases the first column is simply modified to a DWC as a prefraciونator shown in Fig. 4.28a and 4.28c, prefractionated multi-feeds can be formed and enter the subsequent column. These cases can be a safe and cost effective retrofit option, which will be discussed in Sect. 4.7.

4.6.3.1 Initial Structure Design

A shortcut design procedure based on a conventional column configuration was used for the initial design of a DWC structure (Fig. 4.30) (Triantafyllou and Smith, 1992; Long et al., 2010; Lee et al., 2016). The first column in this conventional configuration corresponds to the prefractionator section in the DWC, while the rectifying section of the second column and the stripping section of the third represent the top and bottom sections of the DWC, respectively. The stripping section of the second column and the rectifying section of the third column are equivalent to the dividing wall section of the DWC. A combination of the bottom stream from the second column and the top stream from the third column represents the side stream of the DWC. Consequently, four sections including a prefractionator section for the feed mixture, top and bottom sections above and below the dividing wall section, and a dividing wall section form a DWC. The initial estimates of the DWC structure can be implemented easily by applying a conventional shortcut design method based on this structural similarity.

For example, the structure of each column can be determined using the well-known Fenske-Underwood-Gilliland (FUG) method (Long and Lee, 2013b) as follows:

$$\frac{N-N_m}{N+1} = 0.75\left[1-\left(\frac{R-R_m}{R+1}\right)^{0.5688}\right] \tag{4.3}$$

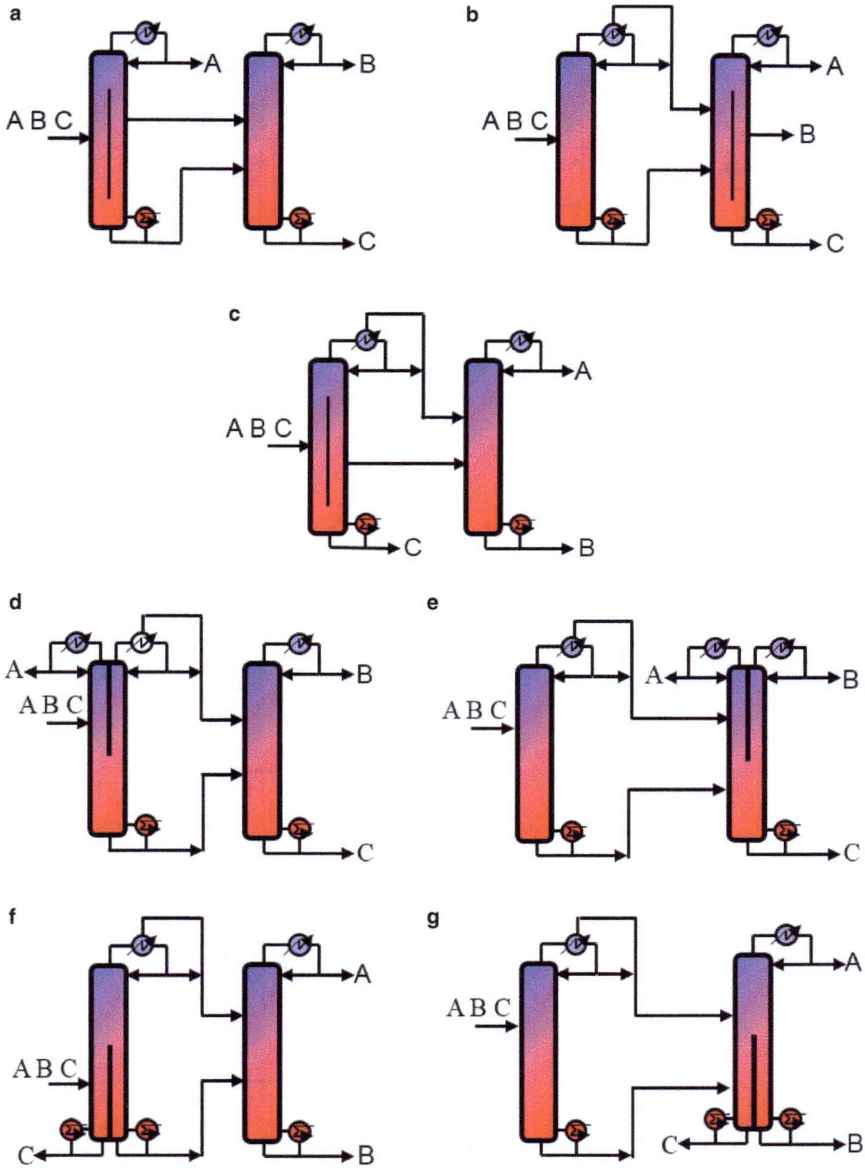

Fig. 4.28 Schematic diagram of promising configurations for the retrofit using DWC

where the minimum number of theoretical stages at the total reflux (N_m) and the minimum reflux for an infinite number of theoretical equilibrium stages (R_m) are estimated using the Fenske equation (Fenske, 1932) and Underwood equation (Underwood, 1948), respectively. The actual reflux ratio (R) is normally chosen to be 1.2 R_m (Rangaiah et al., 2009).

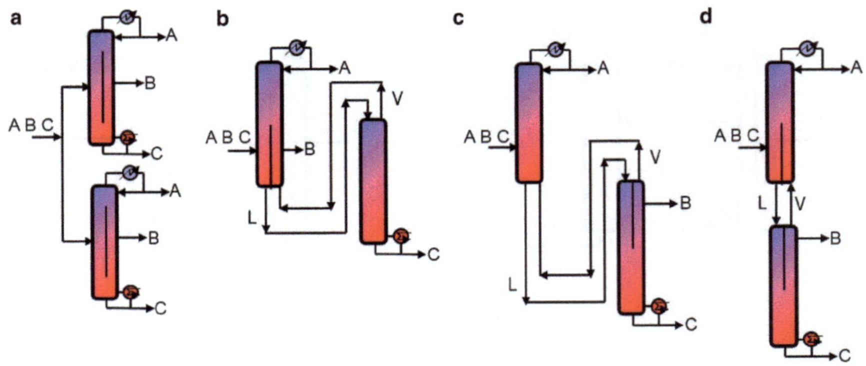

Fig. 4.29 Schematic diagram of promising configurations for the retrofit using DWC

Fig. 4.30 Schematic diagram of three-column distillation system

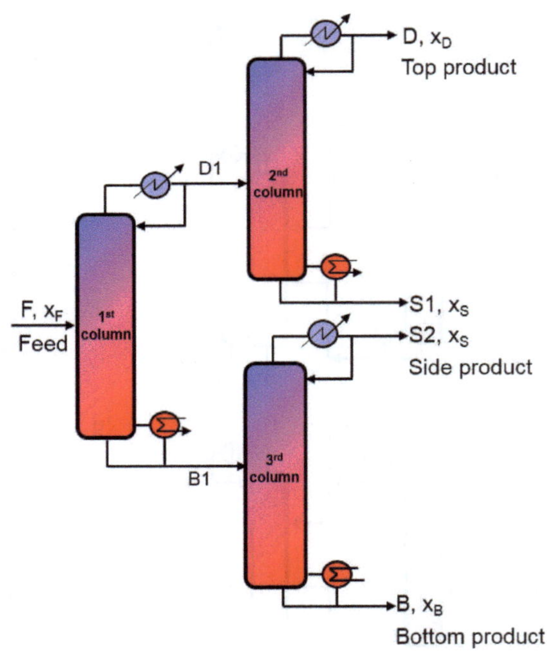

The feed tray location is determined by assuming that the relative feed location is constant as the reflux ratio is changed from the total reflux to a finite value (Wankat, 2012):

$$\frac{N_F}{N} = \frac{N_{F,m}}{N_m} \tag{4.4}$$

where $N_{F,m}$ is the feed stage at the total reflux estimated using the Fenske equation as follows:

$$N_{F,m} = \frac{\ln\left[\left(\frac{x_{D,LK}}{x_{D,HK}}\right) \Big/ \left(\frac{x_{F,LK}}{x_{F,HK}}\right)\right]}{\ln \alpha_{LK-HK}} \tag{4.5}$$

4.6.3.2 Rigorous Simulation

A simulation of a DWC based on the equivalent Petlyuk column has two recycle loops that requires two recycle blocks (*R*1 and *R*2), as shown in Fig. 4.31. The prefractionator column (*Pre*) has no reboiler and condenser, which can be simulated using an absorber column. This column has two recycle streams, which includes a recycled vapor stream to a prefractionator (*Vapor pre in*) in the bottom, and the recycled liquid stream to prefractionator (*Liquid pre in*) in the top. These two streams require the initial estimates to solve the *Pre* column. A distillation unit for the main column (*Main*), the material streams, *Feed*, *Vapor pre out*, *Liquid pre out*, *Liquid main out*, *Vapor main out*, *Top*, *Side*, and *Bottom* for *Main* column are added to the flow sheet. Two recycle blocks need to be added to the flow sheet: one recycle block *R1* for the *Liquid main out* and another *R2* for *Vapor main out* to initialize the recycle stream. Once the *Pre* column is solved, the *Main* column is simulated with an estimated the number of stages and the stage location from shortcut calculation.

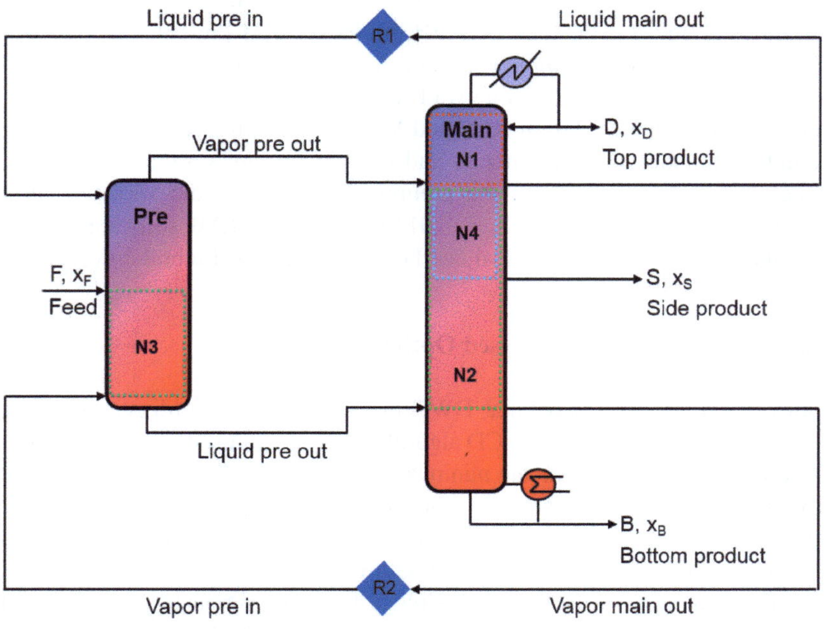

Fig. 4.31 Schematic diagram of Petlyuk column with main design variables

The *Main* column has five degrees of freedom, so five specifications should be provided. The purity and recovery of the main product are normally used as specifications. The flow rate of *Liquid main out* and *Vapor main out* are also considered as specifications. The streams *Liquid main out* and *Vapor main out* are calculated and available for further calculations. Consequently, the complete flow sheet in Fig. 4.31 can converge due to the two recycle blocks. If the initial estimates obtained from the shortcut simulation do not lead to the convergence of a rigorous simulation, small changes in these estimates need to be attempted to achieve convergence.

Another simulation sequence includes first solving the *Main* column by specifying two inlet streams including *Vapor pre out* and *Liquid pre out* that are based on the shortcut design results. Next, the *Pre* is simulated with two recycle blocks located in the *Vapor pre out* and *Liquid pre out* streams.

4.6.3.3 RSM Based Optimization

Because of the large number of mixed variables associated with the DWC design, a statistical approach was found to be a powerful tool for determining the effects of the factors and their interactions (Long and Lee, 2012b). Identifying the most cost effective solution within the constraints of a particular system is the aim of any process design (Kolbe and Wenzel, 2004). In particular, the RSM was used to optimize the structure of the retrofitted DWC (as shown in Fig. 4.32). After initially setting the DWC structure, the main design variables, including the internal vapor (FV) and liquid (FL) flows to the prefractionator, and the number of trays in the top (N1), middle (N2), feed stripping (N3), and side product (N4) sections, need to be optimized (Long and Lee, 2013b). Figure 4.31 presents the main design variables of the DWC. A simplification of the assumption in that the same number of trays in each side of the dividing wall section is often employed but is not mandatory. A Box-Behnken design (Box and Wilson, 1951) under the RSM is used to determine how the variables interact and to optimize the system in terms of the reboiler duty after determining the preliminary ranges of the variables through single-factor testing (Long and Lee, 2013b). The MINITAB software is used to obtain the second order polynomial coefficients and optimal reboiler duty.

4.6.3.4 Coordinate-Decent Based Optimization

Another effective way for optimal DWC design is to use coordinate decent (CD) methodology (Long et al., 2016). CD algorithms solve optimization problems by successively performing approximate minimization along coordinate directions (Wright, 2015). They are iterative methods in which each iteration is obtained by fixing most components of the variable vector x at their values from the current iteration, and approximately minimizing the objective with respect to the remaining components. The main advantages of the CD methodology lie in the simplicity of each iteration, simple implementation, and high efficiency (Long et al., 2016). Thus, this methodology can be suitable for the optimization of highly nonlinear and complex system. The CD can be implemented in the MS Visual basic application and connected to the

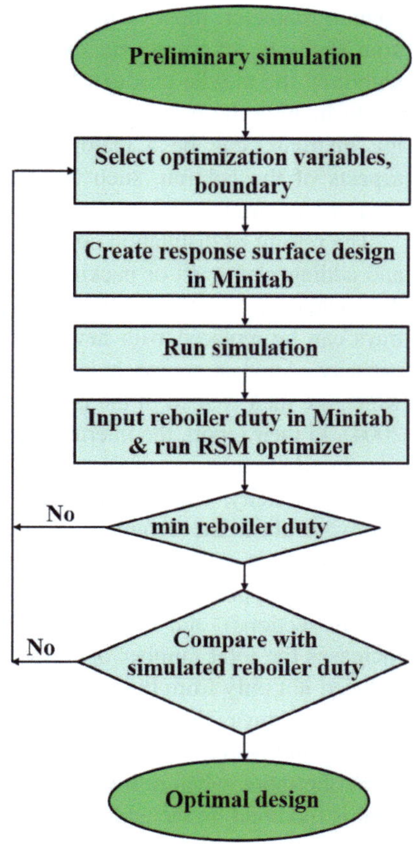

Fig. 4.32 Optimization procedure by the RSM

Hysys model via the MS Excel platform. Excel worksheets and Excel Macro were employed to interface and calculate the objectives as well as implement the optimization algorithm. The CD methodology may lead to a local optimum, and thus a grid search can be first utilized to generate several starting points to avoiding sticking to a local optimum. This approach combining a mathematical programing technique and a statistical method is expected to find the gloal optimum. In addition, box search can be considered to overlook a more promising solution for optimal retrofit.

4.6.4 Column Modification and Hardware

4.6.4.1 Column Modification

Several studies have examined the application of DWCs in a retrofit (Kolbe and Wenzel, 2004; Slade et al., 2006; Premkumar and Rangaiah, 2009; Long et al., 2010; Long and Lee, 2011). They reported that the DWC can be used to save energy and cost. On the other hand, there can be practical difficulties (Amminudin and Smith, 2001). In addition to existing column constraints, such as a fixed column diameter, most of these concerns focused on the need to modify the column shells to accommodate the dividing wall, and have another nozzle fitted to withdraw the middle product. Such modifications to install a dividing wall, which might involve the removal of the tray support rings and the replacement of existing internals, can be a major undertaking. In addition, the mechanical design aspects of the column, such as mechanical stress, are affected by changing the feed or side-draw locations.

The retrofit of distillation columns to DWCs includes removing the existing trays and adding new trays or packing (if necessary) as well as a new dividing wall into the middle section (Long and Lee, 2013b). This might take time but the existing trays can be replaced with new high performance trays or packing to increase the capacity. Another approach is to cut the existing column shell into three pieces, which are then welded in the new prefabricated column section (Kolbe and Wenzel, 2004). In their report, the internals for the new section were installed and leveled on-site. In addition, some of the pump sets were replaced to match the new debottlenecking capacity. Interestingly, they claimed that the modification is planned in such a way to allow the complete reconstruction from the "ready for opening" to the "ready for feed-in" stage in a plant shutdown period of only 10 days.

Another approach is to cut the existing column shell into two pieces (top and bottom sections), and then weld the prefabricated new middle section in, which increases the total number of trays. By carrying out this approach, energy can be reduced not only from the effects of DWC, but also the effect of an increase in the number of trays. Note that the heat losses from the shell and heat transfer across the dividing wall can be ignored in most practice.

Furthermore, when retrofitting two columns to a DWC, the selected, existing column is modified by the addition of a middle section, which houses the dividing wall (Premkumar and Rangaiah, 2009). This will be cut on-site into two portions to form the top and bottom sections of the retrofitted DWC. The middle section of the retrofitted DWC is shop fabricated and would be available before the actual on-site retrofit begins. This section is first welded to the top of the bottom section, and the removed top portion is then welded to the top of the middle section to complete the column. However, a column is not utilized in the proposed solution. Instead of that, the dividing wall is added into two columns for the purpose of complete utilization and arranged in a parallel sequence (as shown in Fig. 4.29a) (Long and Lee, 2014)

When the middle section is still used and the optimized feed and side streams are different from the existing SSC's, it needs to consider keeping their nozzles constant instead of moving to the optimized location, which required some modifications. Although the optimal point provides better operating cost savings, keeping the same location of the feed, and side stream might be preferred because of the fewer modifications required (Long and Lee, 2013b). The trade-off between the operating cost and investment needs to be considered in a retrofit design. Therefore, although the DWC might initially be the best option for retrofit projects, these concerns must be addressed fully during a retrofit design. These modifications might require significant plant downtime, which leads to a loss of production and an interruption of the product supply to the customers.

In the DWC, a dividing wall can be implemented normally by welding a metal plate along the inner walls of the column (Dejanović et al., 2010). On the other hand, the welding job might be very difficult for a small column and it also adds strength to the column, which leads to additional mechanical and thermal stress. To remedy this problem, a partially welded or non-welded dividing wall system can be considered depending on the effects of the internal flow leaks during operation (Dejanović et al., 2010).

All these modifications of existing columns to a DWCs incur significant plant downtime (Long et al., 2010). In many retrofit projects, downtime is the largest economic factor, and a significant amount can lead to a loss of production, an interruption in product supply, as well as a disruption of the other downstream processes, where plants depend on each other (Amminudin and Smith, 2001). On the other hand, chemical/petrochemical plants normally have an annual maintenance period of 10–14 days for major repairs and replacement, where the retrofitting discussed above can be accomplished within this short maintenance period (Premkumar and Rangaiah, 2009). Therefore, the production loss can be negligible during the retrofitting time. In some retrofit projects, replacing a new column is more cost-effective than modifying an existing column body due to the tight turnaround schedule and high labor cost for column modification.

4.6.4.2 Hardware

Two important devices in the DWC are an internal reflux splitter and an internal vapor splitter to properly distribute the liquid and vapor coming from the column overhead and bottom into sections divided by a wall, respectively. The energy efficiency of a DWC can be affected by small deviations in the internal flows.

Several types of reflux splitters are available in commercial DWC implementations. The moving bucket type reflux splitter reported by Montz as a first generation reflux splitting system provides a simple and more precise solution (Asprion and Kaibel, 2010). This splitter consists of a casing that is subdivided into three chambers (Fig. 4.33a). The split ratio is controlled by adjusting the time period of the hollow inner bucket onto the pre-fractionation or the main column chamber. The main potential problem is the moving elements that can be damaged during operation by the constant movement of the mechanical parts. For a more reliable and robust liquid splitting operation, a manual type reflux splitter with no moving parts (Fig. 4.33b) was devised (Lee et al., 2010) and implemented successfully on several commercial DWCs. The latest development in the reflux splitter system was made by a simple modification of the conventional distributor and collector tray (Lee and Kim, 2012). The in-the-column type reflux splitter (Fig. 4.33c) consists of a modified collector tray, a collector box, and two transfer lines. This type of splitter is simpler and cheaper, and can reduce the column height. These different types of internal reflux splitters need to be chosen depending on the retrofit case based on their features listed in Table 4.1.

In most existing DWCs, vapor splitting is fixed in the design stage by a specific fixed open area for vapor flow due to its practical limitation in industrial

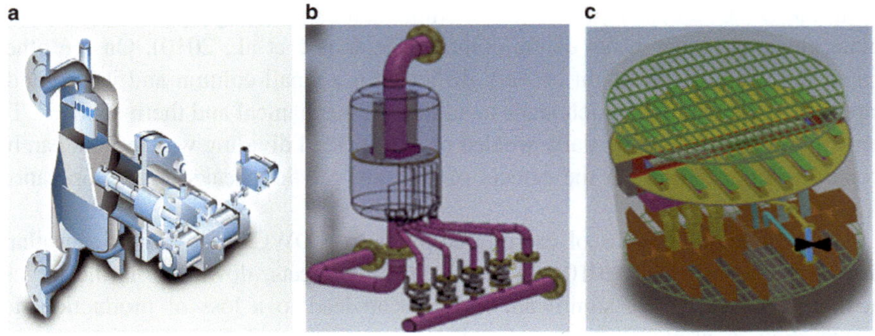

Fig. 4.33 Various internal reflux splitters: (a) moving-bucket type (b) manual type (c) in-the-column type (Shin et al., 2013)

Table 4.1 Comparisons of several reflux splitter systems (Shin et al., 2013)

Splitter type	Operating window/ control	Reliability/maintenance	Cost/structure
Moving bucket type	Wide/continuous	Fair/overhaul needed	High/complex
Manual type	Narrow/discrete	Excellent/regular inspection	Moderate/simple
In-the-column type	Narrow/discrete	Excellent/regular inspection	Low/simple

implementation, and thus cannot be adjusted during column operation. Although the use of a vapor split controller has not yet been popular in industry, several studies have been found on the active vapor splitter.

Dwivedi et al. (2012) proposed an adjustable vapor split consisting of two separate vapor valves that were operated by externally placed electrical motors using a rack and pinion assembly. The vapor flow rate through the valve is manipulated by opening and closing a cap that sits on a steel valve seat. UOP patented a vapor control system that blocked upward vapor flow located below the DW and flowing vapor from below the vapor isolation system to both sections of DWC (King and Haas, 2013). Recently, the blade-shape structure inside of each vapor channel was applied to adjust the vapor split in a DWC by changing the blade angle (Huaqiang et al., 2016). The blade is connected to an axle and operated by an externally placed electrical motor. In their system, there are other elements such as distribution plate used to obtain a uniform gas flow, chimney tray, vapor tubes, and V-type caps are used to avoid the liquid flowing into the vapor tubes.

In developing an ideal internal flow-splitting device for DWC application, the following technical requirements should be considered: (1) the device should not include a pump and moving unit, (2) the additional volume contributed by the device should be small, (3) the device should be able to manipulate and control the split ratio in a stable and accurate manner, and (4) the device should be simple, durable, and reliable (Shin et al., 2013). The moving elements of the liquid/vapor splitter can be damaged by continuous movement of the mechanical parts. Therefore, any liquid/vapor splitting equipment requiring an inserted axle and mechanical/electrical moving unit could have potential mechanical trouble and

Fig. 4.34 3D drawing of the active vapor distributor

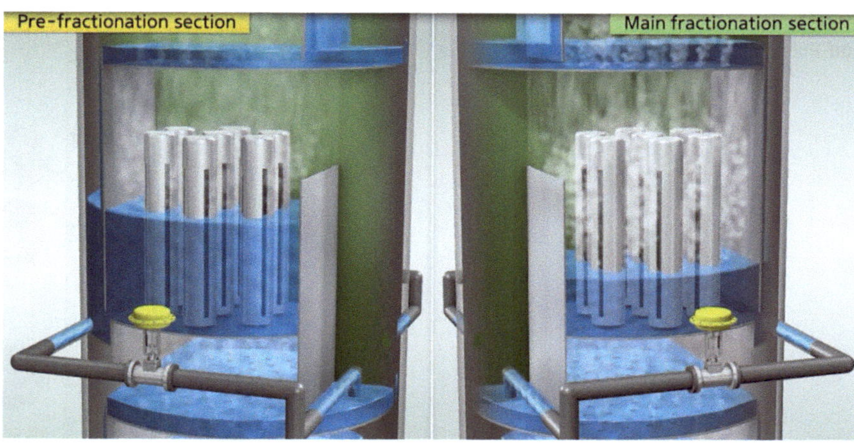

Fig. 4.35 Visualization of the liquid level adjustment of each section using the active vapor distributor

leaking and quaking problems, which would increase the capital and maintenance costs. The concerns associated with these potential problems might limit the real implementation of this type of vapor splitter in DWC.

Recently, a novel active vapor distributor (AVD) (Fig. 4.34) was proposed to address the potential problems associated with a mechanically driven vapor splitter (Kang et al., 2017). It consists of two valves for the liquid level controller and several chimneys (the number and geometry can be varied according to the design) which surrounded by cap that has either continuous or discrete windows as vapor channels. The top side of the caps is closed to circumvent the liquid flowing into the vapor channels.

In the proposed AVD, the friction of a vapor flow path is hydraulically adjusted by changing the liquid level of a modified chimney tray (Fig. 4.35), rather than directly changing the angle or the opening position of a mechanical element inside the column. There is no need for an external motor, a mechanical moving part, or an axle to be inserted in the column; therefore, it can provide a more robust and reliable operation with less maintenance and a simpler structure. It is expected to offer more reliable and robust vapor split during DWC operation. Furthermore, the retrofit of existing DWCs could be done simply by replacing the conventional chimney tray with the proposed chimney tray as an AVD; an additional external device would not be required.

One of main practical barriers in DWC implementation might be the concerns regarding the potentially risky new technologies. A DWC column may also be required to switch from a normal DWC operation mode to a conventional operation mode due to unforeseen circumstances or maintenance. A dual mode DWC configuration (Shin et al., 2011), which can switch its mode of operation by simply varying its valve positions, can be an attractive option to mitigate the concerns of plant workers regarding the potential risks in DWC implementation (Fig. 4.36).

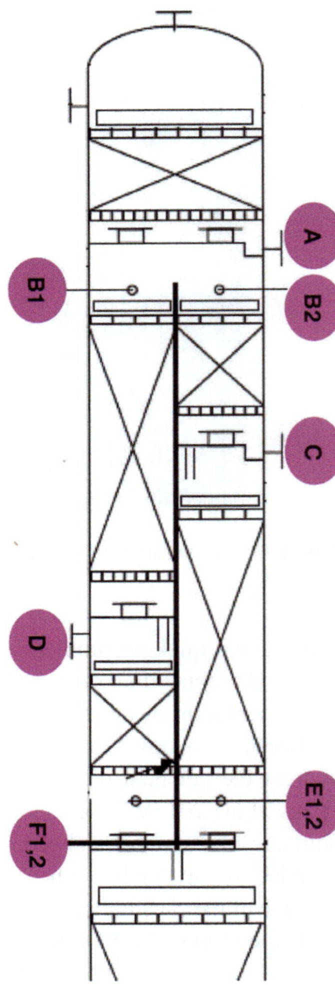

Fig. 4.36 DWC capable of two operation modes (Reprinted from Shin et al., 2011, with permission from Gulf Publishing Company)

4.6.5 *Case Study*

4.6.5.1 Existing Plant Subject to Retrofit Study

Benzene, toluene, and p-xylene (BTX) were separated using conventional distillation column sequence (Premkumar and Rangaiah, 2009; Rangaiah et al., 2009). These authors set the operating pressure to 9.15 bar or 10.13 bar, which caused high heating and cooling duties. This also required a more expensive heating source. Therefore, in this example, the operating pressure of conventional distillation sequence was reduced to 1.01 bar. Figure 4.37a shows the conventional distillation sequence and its current operating conditions (Premkumar and Rangaiah, 2009; Rangaiah et al., 2009). The Peng-Robinson equation of state, which supports the widest range of operating conditions and the widest variety of systems, was used to predict the vapor-liquid equilibrium of these simulations (Aspen Technology, 2009). The base case simulation model showed that the energy consumption of the C-101 and C-202 are 1527 and 1394 KW, respectively.

Because there is a remixing phenomenon in the top section of C-101, it is expected to improve the sequence performance when retrofit these columns using DWCs. Here the configuration in Fig. 4.29a was used to show the benefit of using DWCs in a retrofit. To retrofit the conventional distillation sequences to two parallel DWCs, the main design parameters of the existing columns were checked. The retrofit was carried out with no change in the diameter or the total number of stages of each column to

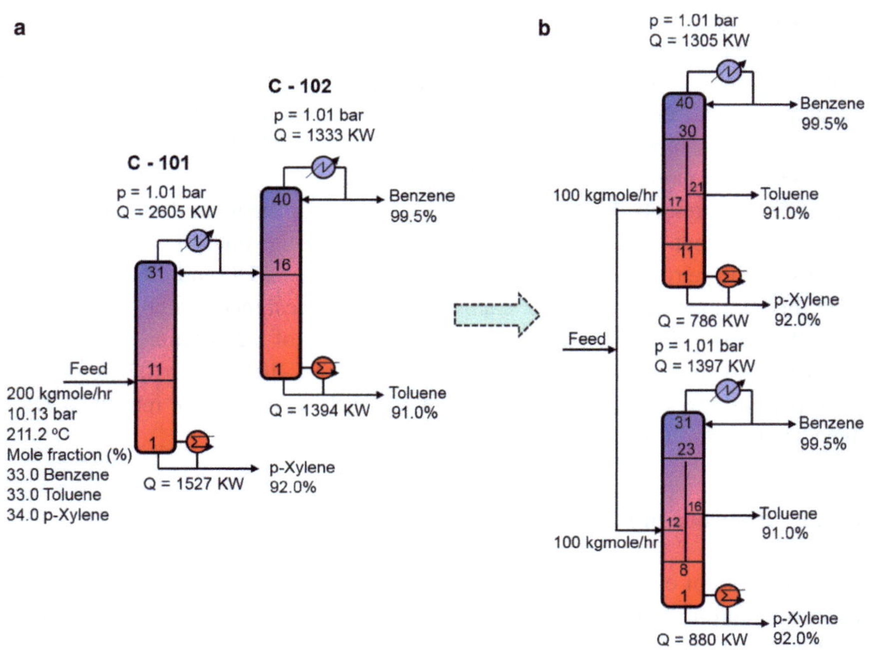

Fig. 4.37 Simplified flow sheet illustrating (a) the existing separation train of two conventional columns for the BTX mixture; (b) retrofitted configuration

highlight the optimal use of the existing columns. Furthermore, the existing reboilers and condensers should be checked for reusability with minimal modifications. As there are two parallel DWCs, the feed requires splitting. After optimizing the DWCs structure and operating conditions, the feed ratio is then optimized to maximize the energy savings. Normally, the retrofitted DWC with a larger number of trays requires less energy than the retrofitted DWC with a smaller number of trays. Therefore, that column will receive the maximum feed flow rate; the remaining feed will be handled by another retrofitted DWC. In particular, the retrofitted DWC from C-102 is considered to have a larger feed flow rate. For an initial estimation of the structure of a DWC, the feed is firstly assumed to split with a ratio of 1:1.

4.6.5.2 Initial Structure Design

A shortcut design method was used to estimate the initial DWC structure. A conventional three column configuration in Fig. 4.30 was used to determine the initial number of trays of each section of DWC as well as the feed and side stream location by assuming the light key in bottom, heavy key in the distillate and specifying the actual reflux ratio. In particular, the first column was simulated assuming a mole fraction of 0.0165 of the light component, benzene, in the bottom (*B1*) and 0.0125 of the heavy component, p-xylene, in top (*D1*). The second column was then simulated by specifying the light key in bottom benzene and heavy key in distillate toluene are 0.05 and 0.005 mole fraction, respectively, to achieve the required purity. Similarly, the third column was simulated using the light key in the bottom, toluene, and the heavy component key component, p-xylene were 0.08 and 0.0419 mole fraction, respectively. The specifications of benzene in *B1* and p-xylene in *D1* of the first column were varied by trial and error to achieve an almost equal composition of *S1* and *S2*. The actual reflux ratio (R) for the three columns was chosen as 1.2 R_m. Note that in the three-column model, the middle product is represented by the addition of two material streams, namely *S1* and *S2* in Fig. 4.30.

Figure 4.38 presents the resulting estimated values of the main design parameters. From this shortcut model, the number of stages for the bottom, middle and top sections of the retrofitted DWC were calculated to be 7, 18, and 15, respectively. After the number of stages for each section was designed initially, they were adjusted relatively to satisfy the total number of stages of the column being same as the existing columns. Because the number of stages of C-101 and C-102 were 31 and 40, respectively, only the number of stages for each section in the retrofitted DWC from C-101 was adjusted.

4.6.5.3 Rigorous Simulation

Once the shortcut estimates are completed, the DWC needs to be simulated rigorously to obtain accurate results. The approaches for a rigorous simulation involve fewer assumptions that are more realistic than the FUG equations (Premkumar and Rangaiah, 2009). The rigorous simulation used in this case study was based on the

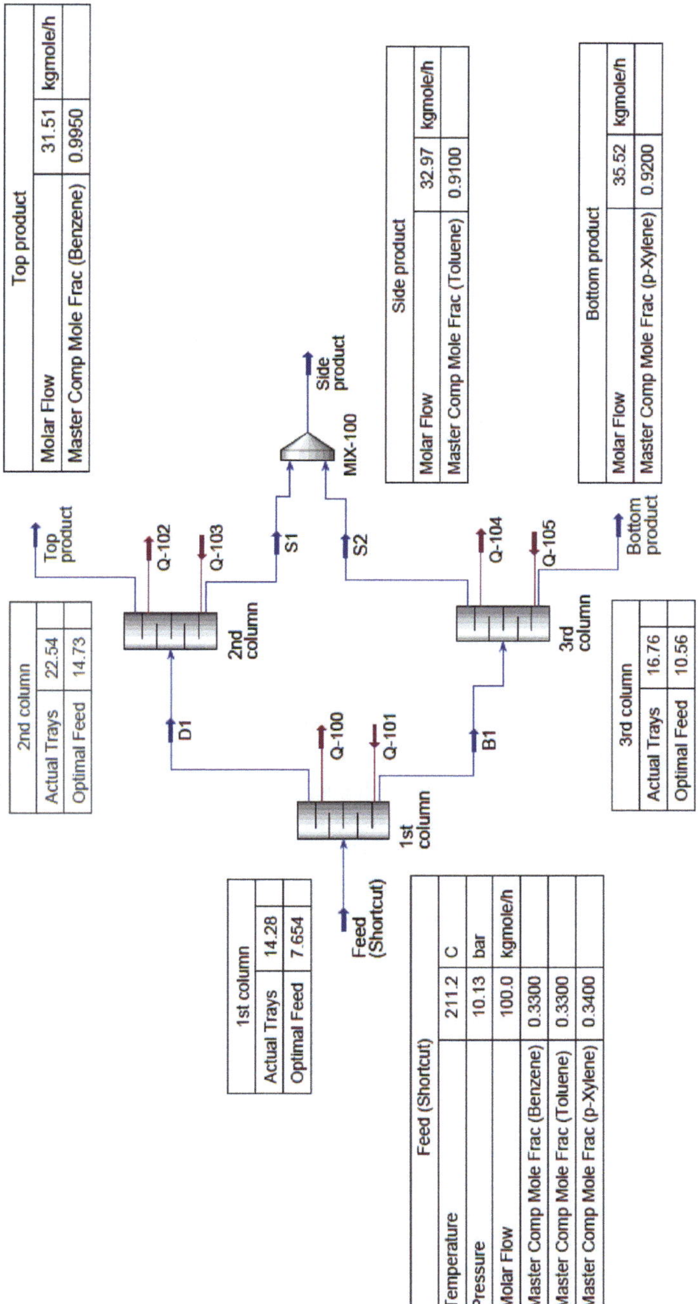

Fig. 4.38 Three-column shortcut simulation for initial design of DWC

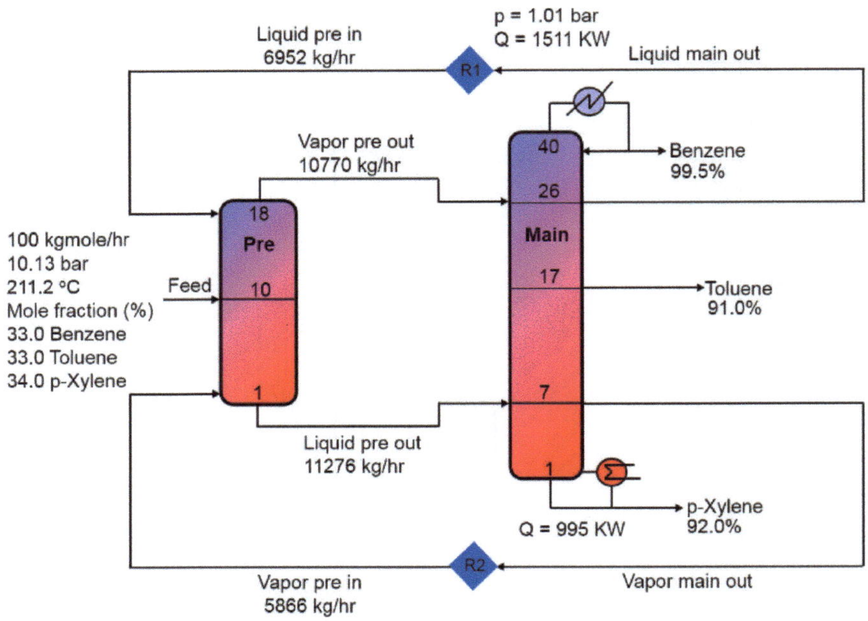

Fig. 4.39 Rigorous simulation of retrofitted DWC configuration

equilibrium-stage model. The simulation model of a DWC based on the equivalent Petlyuk column has two recycle blocks (*R1* and *R2*), as shown in Fig. 4.39. Note that in this figure, the stage is numbered from the bottom up.

In this case, the purity of benzene, toluene, and p-xylene in the distillate, side, and bottom stream, respectively, were used as three specifications. The flow rate of *Liquid main out* and *Vapor main out* should be specified as the remaining two specifications. The mass and energy balance of the base case were obtained by activating this configuration (Fig. 4.39).

4.6.5.4 Structure Optimization Using RSM

After the design using a shortcut and rigorous simulation, the main design variables associated with the column structure were optimized by the RSM using a simplified assumption of the same number of trays in each side of the dividing wall section. The preliminary ranges of the variables need to be determined by single-factor testing. The variables interaction and optimization of the system in terms of the reboiler duty were found by RSM with the Box-Behnken design. Table 4.2 lists the factors and levels used in this case study. Factors are defined as design variables that affect the objective. In the RSM, each factor was defined on a coded scale from -1 to +1 (the low and high levels of factors).

Table 4.2 Coded levels of factors

Factors	Levels		
	−1	0	1
Top section (N_1)	6	10	14
Middle section (N_2)	18	23	18
Side product location (N_3)	4	7	10
Feed stripping location (N_4)	6	9	12

To optimize these four parameters associated with the DWC structure, 27 simulations were run. The resulting second-order polynomial model was as follows:

$$\begin{aligned} Y = {} & 785.00 - 32.99N_1 - 1.63N_2 + 13.75N_3 + 7.74N_4 + 59.30N_1^2 + 72.24N_2^2 \\ & + 30.12N_3^2 + 36.25N_4^2 + 38.08N_1N_2 - 3.03N_1N_3 - 3.83N_1N_4 \\ & - 13.63N_2N_3 - 14.90N_2N_4 + 2.15N_3N_4 \end{aligned} \tag{4.6}$$

As expressed in Eq. (4.6), in this case, the top section had a significantly greater influence on the reboiler duty than the side product location, followed by the feed stripping location and the middle section. The quality of the polynomial model was expressed by the coefficient of determination, that is, R^2. Based on the analysis results from MINITAB, the model showed a high R^2 value of 0.98. The value of R^2 is close to 1.0, which indicates a strong correlation between the simulated and predicted values. The relationship between the independent variables and the response is well explained by the regression model.

The minimum reboiler duty was found at the coded levels of the number of trays in the top, middle, feed stripping, and side product sections of 0.29, −0.09, −0.23, and −0.11, respectively (Fig. 4.40). Under these conditions, the minimum reboiler duty was predicted to be 778 KW. The natural values of these valuables can be derived from the coded levels. FV and FL were then optimized to minimize the reboiler duty, whilst maintaining adequate product recovery and purity. The result from a rigorous simulation also showed that the retrofitted DWC requires a reboiler duty of 786 KW, which shows good agreement with the result predicted by the RSM.

As discussed above, the exiting conventional column sequence was retrofitted to configuration in Fig. 4.29a, which has two parallel DWC. Because the retrofitted DWC from C-102 has larger number of trays (40 trays) compared to the retrofitted DWC from C-101 (31 trays), it will receive the maximum feed flow rate. From a rigorous simulation, the maximum flooding in the retrofitted DWC from C-102 is 84.1% with a feed split ratio of 1:1. Therefore, the remaining 100 kgmole/hr of feed is separated in the retrofitted DWC from C-101.

Figure 4.37b shows a simplified retrofitted configuration. Compared to the existing conventional distillation sequence, the use of the proposed sequence can reduce condenser duty and reboiler duty significantly by 31.4% and 43.0%, respectively. The decrease in CO_2 emissions that is linked to the lower energy

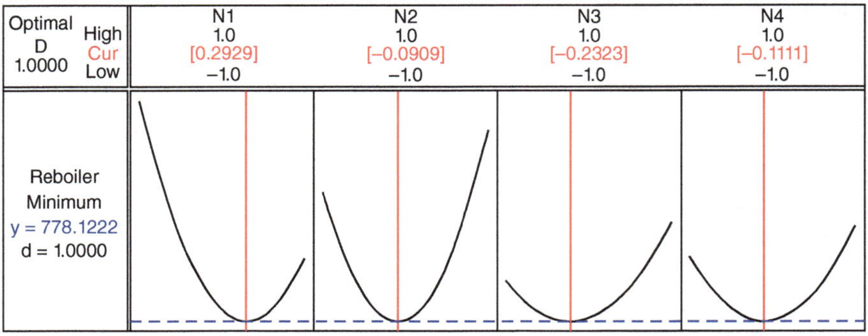

Fig. 4.40 Optimization plot by RSM

requirement is another key advantage with respect to the retrofit of conventional two-column sequence to DWCs.

4.6.5.5 Structure Optimization Using CD Methodology

The CD methodology was implemented in the MS Visual basic application and connected to the Hysys model via the MS Excel platform. The optimization results from the CD methodology represent the reboiler duty saving of 43.1%, which is similar to the simulated result estimated from the RSM, compared to the existing sequence. The optimal DWC structure and operating conditions are determined in a practical and effective manner, and simple to implement. This makes it suitable for industrial uses. It should be noted that maximum reboiler duty saving of 43.0% is achieved while using the DWC with 68 trays, which is much higher than that in the proposed sequence (Premkumar and Rangaiah, 2009).

4.6.5.6 Column Hydraulics

To modify a conventional distillation column, a wall was installed in the middle section with no change in the diameter and the total number of trays. The column hydraulics and flooding need to be checked carefully during a retrofit design. To determine the maximum flooding of the columns considered, the rating mode was simulated using the column internal specifications, such as the type of trays, column diameter, tray spacing, and number of passes (Long and Lee, 2011). The results from rigorous simulations showed that the maximum flooding in retrofitted DWC from C-102 was 84.1% while that in retrofitted DWC from C-101 decreased from 83.0% to 46.2% through a retrofit to the DWC, which suggests that the plant production capacity can be increased after a retrofit. The simple payback period was found attractively in only 5.1 months because of high energy costs saving and the re-use of most of the existing equipment.

4.7 Dividing Wall Prefractionator Configuration

4.7.1 Introduction

As shown in Fig. 4.3a and 4.3b, the first column can be retrofitted to a SSC to utilize the remixing problem, and then the side stream also is fed into the second column. The first column plays as a prefractionator, which can improve the separation efficiency of subsequent column. However, since the prefractionator itself requires additional energy to generate the multi-feed streams, overall energy efficiency improvement in a retrofit project using prefractionation idea mainly depends on how effectively the prefractionator can be designed in terms of cost and energy efficiency. To address this issue, the prefractionation was implemented by modifying the existing column to DWC by installing a dividing wall inside the column to form dividing wall prefractionator configurations (DWPC) as shown in Fig. 4.28a and 4.28c. They are expected to provide multi-feed streams to the second column in a most efficient way by fully utilizing the advantages of DWC. Moreover, the second column can take an advantage of improved separation efficiency from the prefractionated multi-feed streams.

4.7.2 Case Study

4.7.2.1 Existing Conventional Sequence Column

In this example, a retrofit of the NGL sequence in Sect. 4.2 was carried out. The purpose of a retrofit project is to increase the energy efficiency with the following assumptions:

- The existing columns are already operating with the highest performance internals,
- All columns are fully utilized,
- The recovery and purity of all products are kept constant.

4.7.2.2 Proposed Sequence

For a retrofit, the existing debutanizer was modified to the dividing wall prefractionator by installing a vertical wall. As shown in Fig. 4.41, the reboiler duty of the deisobutanizer was significantly reduced up to 21.4% from 6983 KW to 5490 KW while the reboiler duty of the retrofitted debutanizer remained similar level to the existing one. Note that the DWPC can be switched over to the conventional sequence configuration again to continue operation as a contingency owing to unforeseen circumstances. The proposed DWPC enables mutual conversion between the DWC and the conventional column operating mode without shutting down a process while minimizing the operational risks and ensuring a safe retrofit

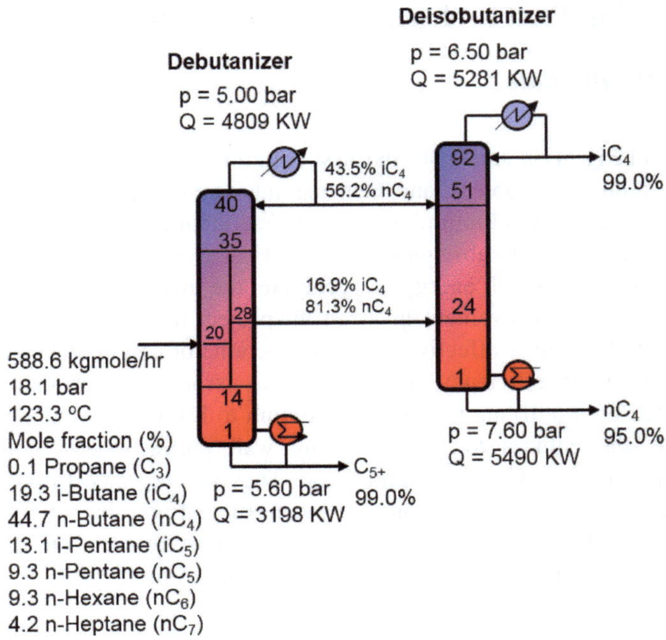

Fig. 4.41 Simplified flowsheet illustrating the retrofitted DWPC

to an existing column (Shin et al., 2011), which is the additional benefit of the proposed DWPC.

4.8 Multi-Effect Configuration

4.8.1 Introduction

As mentioned in Sect. 3.7, a multi-effect distillation system allows a significant reduction in energy consumption, since the condensers and reboilers of different columns can be integrated thermally. Here this method is utilized to improve the performance of multi-column system used to separate multicomponent.

Figure 4.42a shows the configuration of a direct sequence without energy integration, where the bottom product of the first column is the feed to the second column. Using the heat of the top vapor from the first column as a heat source in the reboiler of the second column after adjusting the pressure, the multi-effect direct sequence with forward integration (see Fig. 4.42b) was constructed, where two conventional columns were heat integrated in the same direction as the process flows. The pressure in the reflux drum of the first column was set by

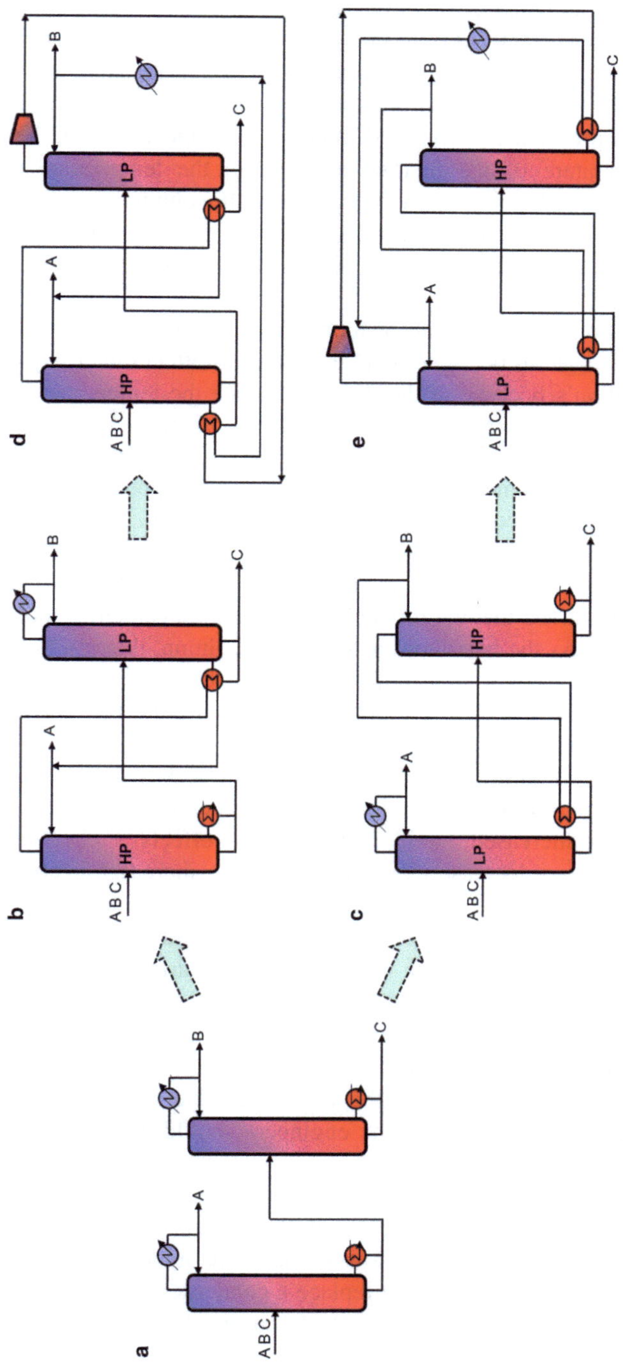

Fig. 4.42 Schematic diagram of the (a) direct sequence, (b) multi-effect direct sequence with forward integration, (c) multi-effect direct sequence with backward integration, (d) multi-effect direct sequence with forward integration combining with MVR, and (e) multi-effect direct sequence with backward integration combining with MVR

the bubble pressure of the overhead distillate at a temperature that was 10°C higher than the bottom temperature of the second column.

In contrast, Fig. 4.42c differed from Fig. 4.42b in the direction of heat integration. The overhead vapor of the second column was used as a heat source to the reboiler of the first column. In this case, the pressure of the second column was set by the bubble point pressure of the overhead distillate at a temperature10°C higher than the bottom temperature of the first column. When the temperature difference between the top of the LP column and the bottom of HP column is small, heat pump assisted LP column can be considered to conserve energy (Fig 4.42d,e). Using the same principle, the multi-effect also can be constructed for an indirect sequence, which is shown in Fig. 4.43.

The pressure level in HP column of multi-effect direct sequence with forward integration may be need to be higher than that of multi-effect direct sequence with backward integration's because in the direct sequence the temperature difference between the first column's top and second column's bottom is larger than that between first column's bottom and second column's top. On the other hand, the multi-effect indirect sequence with forward integration appears more attractive than multi-effect indirect sequence with backward integration.

The forward heat integration arrangement is easier to control compared to the backward heat integration arrangement (Bildea and Dimian, 1999). Engelien and Skogestad (2005) claimed that the forward-integrated arrangement would be easier in terms of start-up. As the heat input is to the first column, this can be started up first, for example, by using a total reflux approach, then when the first column is up and running, it will be relatively easy to start the second up. The backward-integrated arrangement would be more difficult.

On the other hand, although multi-effect distillation has indeed many benefits, it also has some application limitations. The operating pressure in the HP column needs to be increased, having some adverse effects, such as more difficult separation due to a decrease in relative volatility, increased heating cooling temperature, in which a more expensive heat source is needed in the reboiler, and the column shell may need to be thicker to withstand the changed pressure. Therefore, engineers should check the mechanical durability of the existing column shell when the operating pressure is increased in a retrofit. Furthermore, operation and control of these sequences are more difficult compared to the conventional column sequence. Decreasing the pressure in a LP column causes an increase in the cooling medium in the condenser due to a reduction in top temperature difference. In addition, the engineers should be aware of avoiding the use of refrigeration.

The savings can be enhanced using a prefractionator arrangement (Cheng and Luyben, 1985; Emtir, 2001; Engelien and Skogestad, 2005; Long and Lee, 2013c). Figure 4.44 shows simplified flow sheets of the multi-effect prefractionator arrangements. One heat exchanger can replace a condenser in the HP column and a reboiler in the LP column. The multi-effect concept is also extended to sloppy arrangement (Fig. 4.45) for improving the energy efficiency.

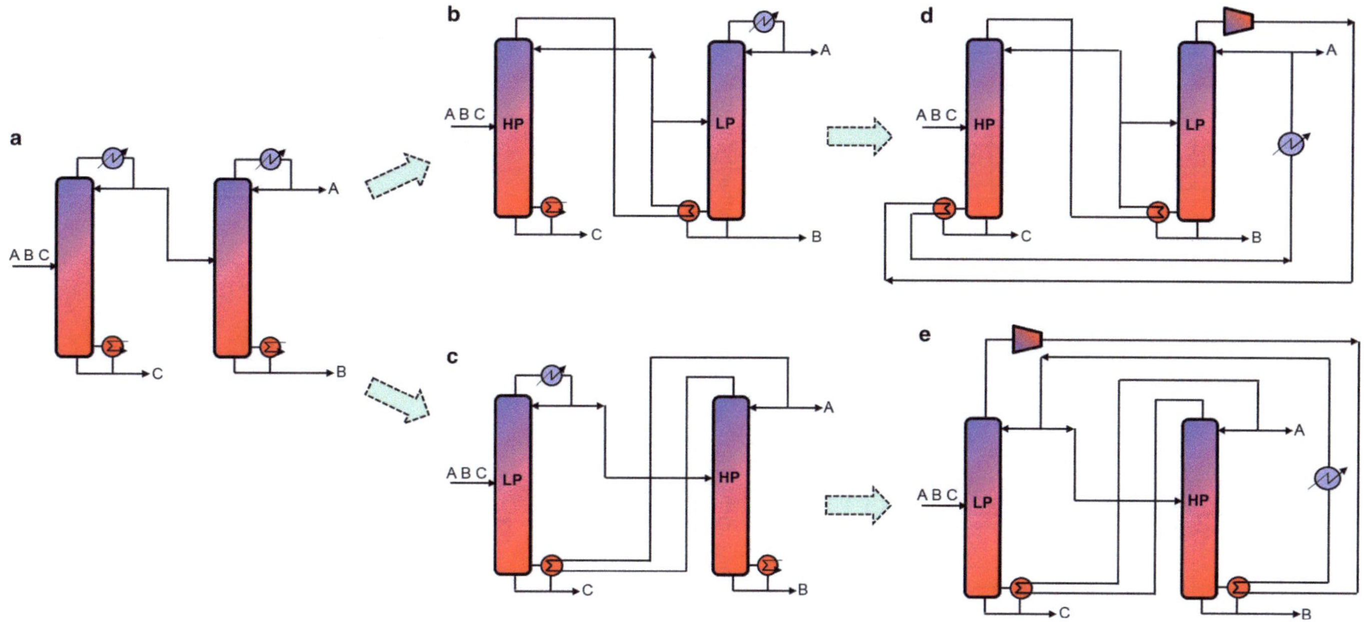

Fig. 4.43 Schematic diagram of the (a) indirect sequence, (b) multi-effect indirect sequence with forward integration, (c) multi-effect indirect sequence with backward integration, (d) multi-effect indirect sequence with forward integration combining with MVR and (e) multi-effect indirect sequence with backward integration combining with MVR

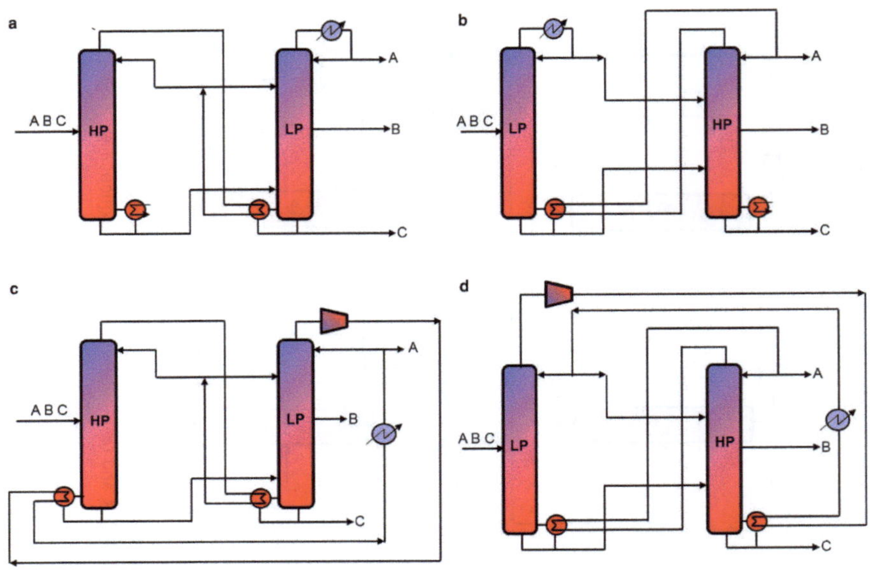

Fig. 4.44 Schematic diagram of the (a) multi-effect prefractionator arrangement with forward integration, (b) multi-effect prefractionator arrangement with backward integration, (c) multi-effect prefractionator arrangement with forward integration combining with MVR, and (d) multi-effect prefractionator arrangement with backward integration combining with MVR

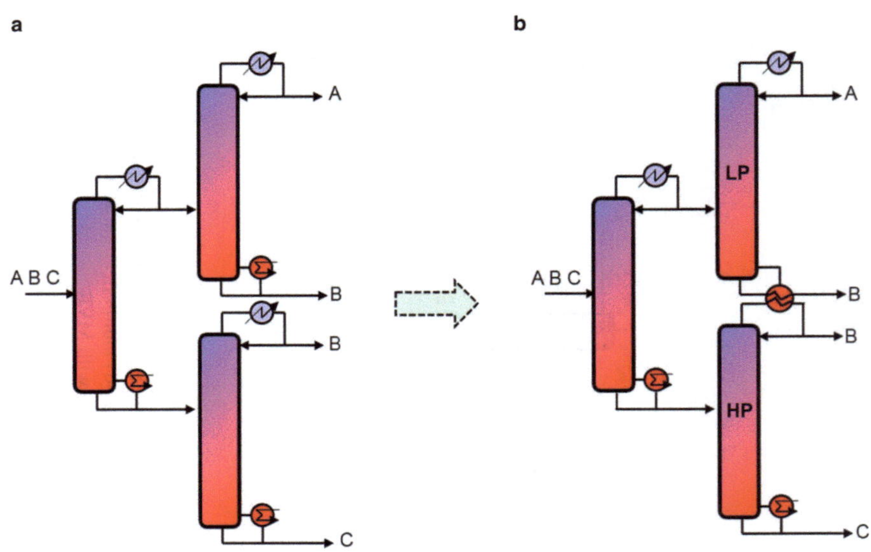

Fig. 4.45 Schematic diagram of the (a) sloppy arrangement and (b) an attractive candidate of a multi-effect sloppy arrangement

4.9 Combination of Internal and External Heat Integration

4.9.1 Introduction

Conventional distillation configurations require both high operating costs and large capital investment, which can be improved in two ways. The first is heat integration, which normally utilizes the heat from the top vapor stream or bottom product (external heat integration). Another is using the thermal coupling concept, which can eliminate reboiler and/or condenser compared to the conventional distillation sequences (internal heat integration). Distillation configurations, which include internal and external heat integration, can reduce the energy requirements significantly.

Furthermore, utilizing only one technique sometimes has limitations due to the small improvement, and/or equipment constraint, which can be enhanced by combining several techniques, which can support each other. Of course, this can cause some problems related to design, optimization, and operation as well as higher investment cost. This section mainly shows some promising combinations of internal and external heat integration to determine the advantages of use these configurations.

4.9.2 TCDS or DWC with Heat Pump or Self-Heat Recuperation Technology

A heat pump and self-heat recuperation are economic ways of conserving energy when the difference in temperature between the overhead and bottom of the column is small and the heat load is high. Therefore, retrofitted TCDSs and retrofitted DWCs that satisfy the above conditions can be considered to add a heat pump and apply self-heat recuperation technology. There are many combinations between retrofitted TCDSs or retrofitted DWCs with heat pump or self-heat recuperation technology. Figure 4.46 presents a schematic diagram of an attractive candidate of a combination of internal and external heat integration. Although the BDWC normally

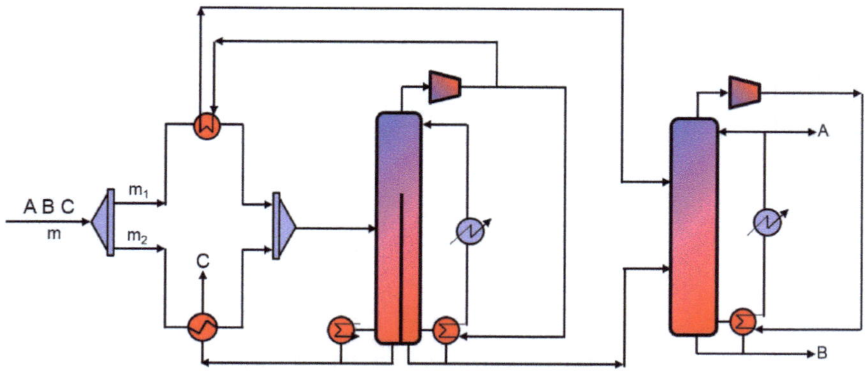

Fig. 4.46 Schematic diagram of an attractive candidate of a combination of an internal and external heat integration

saves less energy than a conventional DWC, it has a smaller temperature difference between the top and right side column bottom (Long and Lee, 2013c). This can provide an opportunity to add a heat pump or apply self-heat recuperation technology. The second column can take an advantage of improved separation efficiency from the prefractionated multi-feed streams. In addition, this configuration enables mutual conversion between the BDWC and the conventional column operating mode without shutting down a process while minimizing the operational risks and ensuring a safe retrofit, which is the additional benefit of the proposed configuration.

As discussed in Sect. 4.4, retrofitting a conventional distillation sequence to a retrofitted TCDS can create a bottleneck in the column. Adding a side reboiler or a side condenser to a retrofitted TCDS can remove the bottleneck problem. Normally a retrofitted TCDS with side rectifier has a bottleneck problem in the bottom section of the main column, which can be solved by adding a side reboiler. The use of a side reboiler can increase the capacity and reduce the temperature difference between the top and side reboiler location, which can use a heat pump to utilize the heat from the top vapor stream. As shown in Fig. 4.47, TCDS was designed by implementing the interconnecting streams (one in the vapor phase and the other one in the liquid phase) between the two columns, which eliminate one reboiler. The vapor from the top of the distillation column can be compressed to a certain pressure (increasing its vapor temperature) and then condensed in the column's side reboiler. This allows the heat of the condensing vapor to assist in vaporization at the middle of the column via a heat exchanger. The top outlet stream must be cooled further before being divided into two streams: one that is recycled back into the column as reflux, and another that is the final top product.

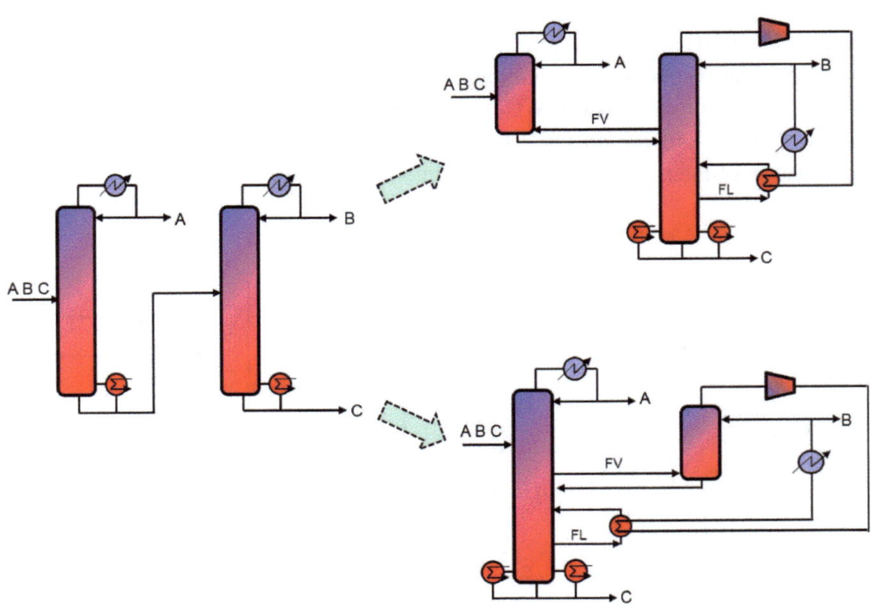

Fig. 4.47 Simplified flow sheet illustrating the retrofitted TCDS with top vapor recompression heat pump and side reboiler

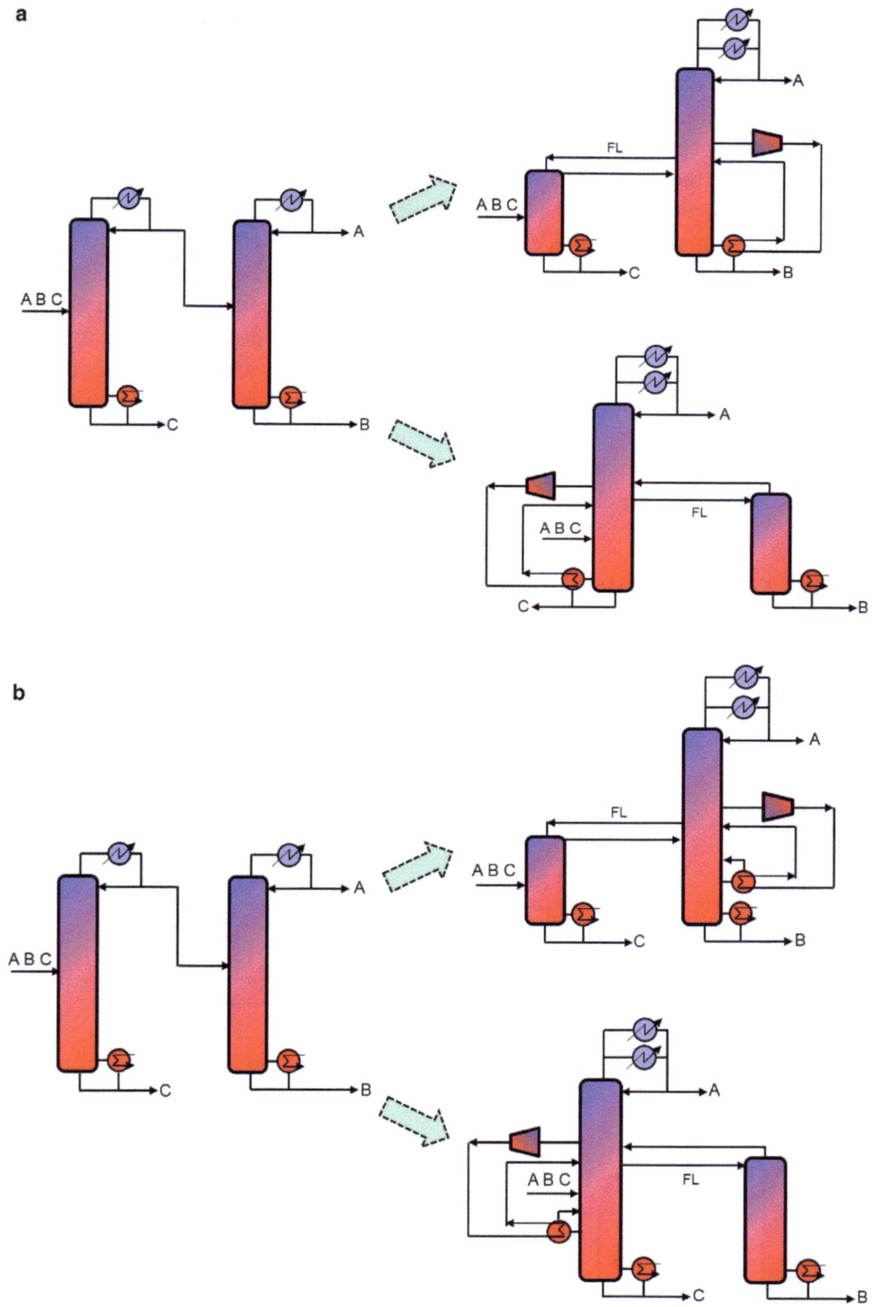

Fig. 4.48 Schematic diagram of the (a) retrofitted HPTCDS and (b) retrofitted HPTCDS with a side reboiler

To solve the same problem when retrofitting to a side stripper, a proposed hybrid process combining a heat pump and thermally coupled distillation sequence (HPTCDS) was employed. Figure 4.48a presents the possibility of re-arranging two existing columns into a HPTCDS. In addition, instead of supplying heat to the reboiler, a heat pump can be used to provide side heating (Fig. 4.48b) to reduce the traffic in the stripping section, which provides more opportunity to increase the capacity. Furthermore, supplying heat to the side reboiler can reduce the temperature difference between the heat pump and side reboiler location, which allows a decrease in heat pump duty. Debottlenecking the existing distillation sequence begins with the development of preliminary designs for complex systems that minimizes the operating cost through the optimization procedures. The HPTCDS was designed using a thermal link in the vapor phase in a conventional indirect sequence, which eliminates the condenser in the conventional scheme and adds a heat pump, which utilizes the heat in the condenser. Modification for the retrofit involves a change in the draw trays for the vapor and liquid streams. More pipe work is also required for the thermal link and re-assignment of the eliminated condenser to the main column.

4.9.3 *TCDS or DWC with Multi-Effect*

The multi-effect concept can be used in the TCDS (Fig. 4.49) and DWC (Fig. 4.50) to improve the energy efficiency. Feed stream is required to split in the retrofit to

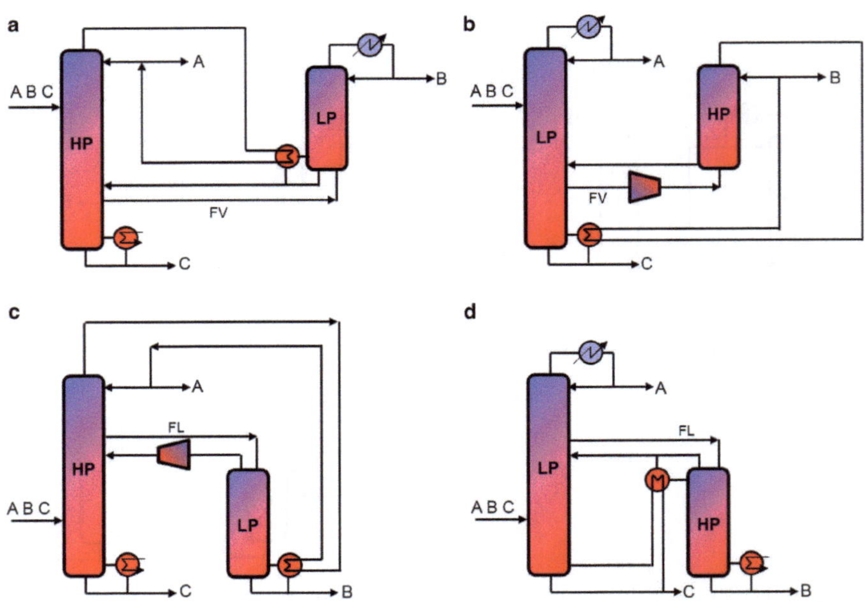

Fig. 4.49 Schematic diagram of the (a) multi-effect TCDS with side rectifier by forward integration, (b) multi-effect TCDS with side rectifier by backward integration, (c) multi-effect TCDS with side stripper by forward integration and (d) multi-effect TCDS with side stripper by backward integration

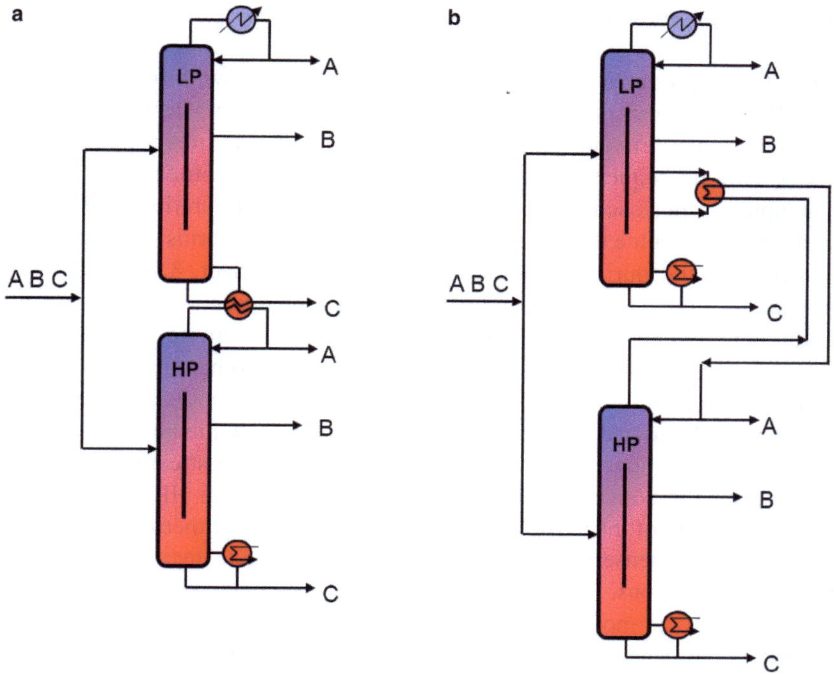

Fig. 4.50 Schematic diagram of the (a) multi-effect DWC and (b) multi-effect DWC with a side reboiler

a multi-effect DWC, which has two parallel DWCs (Fig. 4.50a). Note that the multi-effect heat integration is one of the best examples of proven process integration, whereas DWC is one of the best examples of proven process intensification of distillation process. Their combination is expected to bring a substantial saving after retrofit.

4.9.4 Case Study

4.9.4.1 Existing Conventional Sequence Column

In this example, a retrofit of the NGL sequence in Sect. 4.2 was carried out. The purpose of a retrofit project is to increase the throughput without producing a bottleneck and reduce the energy requirement with the following assumptions:

- The capacity must increase by 10% over the existing one,
- The existing columns are already operating with the highest performance internals,
- No change in the diameter or the total number of stages of each column to highlight the optimal use of the existing columns,
- The recovery and purity of all products are kept constant, and
- The FUA_{max} of all existing columns is 1 or maximum flooding of all existing columns is 85%.

4.9.4.2 Proposed Sequence

With a target production capacity of up to 10%, the new base was simulated using the old process configuration, where all the columns were used to obtain the products. Unfortunately, the results showed that it is impossible to use the existing column sequence because all the columns are bottlenecked when the throughput is increased. Therefore, the HPTCDS configuration in Fig. 4.48a was used to determine the requirements of the product in terms of the productivity, recovery, and purity. In addition, as discussed in Sect. 3.2, the deisobutanizer energy efficiency can be increased using a heat pump operating between the top and bottom with relatively small differences in temperature by an inspection of the CGCC. Therefore, in this example, two heat pump systems are employed accompanying by TCDS. One is side a heat pump used to remove the bottleneck problem and improve the energy efficiency, while another heat pump is installed in the top to utilize the heat from the top vapor stream.

To retrofit the conventional distillation sequences to the HPTCDS, the main design parameters of the existing columns were checked. Furthermore, the existing reboilers and condensers should be checked to determine their reusability with minimal modifications.

To minimize the modifications in a retrofit, the number of trays of all columns was fixed to the same number in the conventional distillation columns. After a preliminary design for the HPTCDS column, the main design variables, such as the liquid flow (FL) to the side stripper, liquid (N1) and feed (N2) stream locations, were then optimized. The RSM was used to examine the effects on the operating cost savings. After determining the preliminary ranges of the variables through single-factor testing, a central composite design was used to determine how the variables interact as well as to optimize the system and achieve operating cost savings. Thirteen simulations were run to optimize the two parameters of the HPTCDS structure.

Figure 4.51 shows a contour plot of the interaction between the main design variables, N1 and N2. The maximum operating cost saving by the proposed sequence was predicted to be 51.7% at the coded levels of the number of trays in the vapor stream and feed sections of −0.41 and −0.13, respectively. The natural values of the variables can be derived from the coded levels. The liquid flow to the side stripper was then optimized to maximize the operating cost, while maintaining sufficient product purity. Figure 4.52 presents a simplified flow sheet illustrating the proposed HPTCDS. Compared to the conventional column sequence, the use of a HPTCDS can reduce the condenser duty and reboiler duty by up to 49.7% and 64.9%, respectively. This resulted in an operating cost saving of 51.7%, which is in good agreement with the results predicted by the RSM. Note that the reboiler duty and operating cost savings generated from the HPTCDS without the heat pump system in the top were 40.1% and 36.3%, respectively. No steam was used in the main column of the prosed sequence by installing two new compressors and two heat exchangers to utilize the heat from the top and intermediate vapor streams. In this case, the simple payback period was found to be 25.1 months.

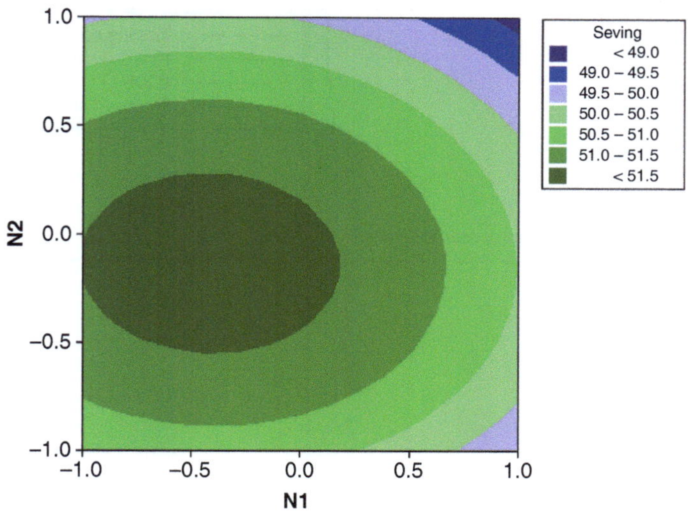

Fig. 4.51 Contour plot of interaction between N1 and N2

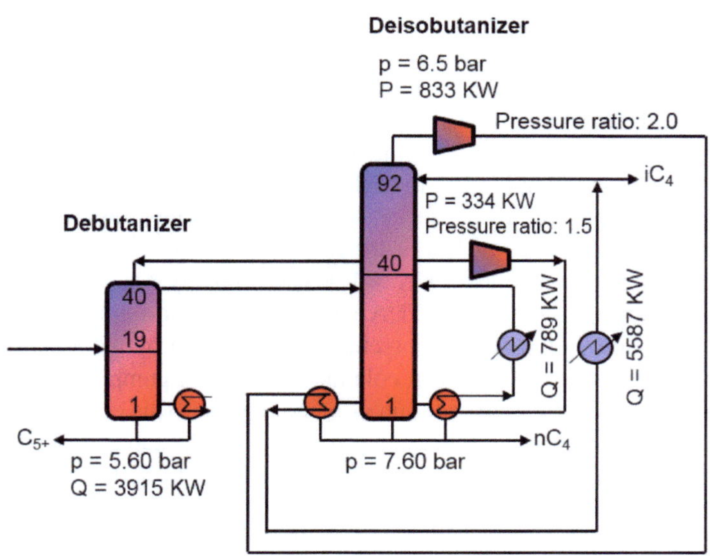

Fig. 4.52 Simplified flow sheet illustrating the proposed sequence

Figure 4.53 shows the FUA curve, which provides graphical information on the hydraulic performance. The FUA varies according to the stage, with all stages requiring an area larger than the available area (FUA>1) when increasing the capacity up by 10%. This indicates that flooding occurred in the entire column after increasing capacity. This bottleneck problem cannot be removed using only a heat pump. This is because the use of a heat pump cannot alter the vapor traffic in

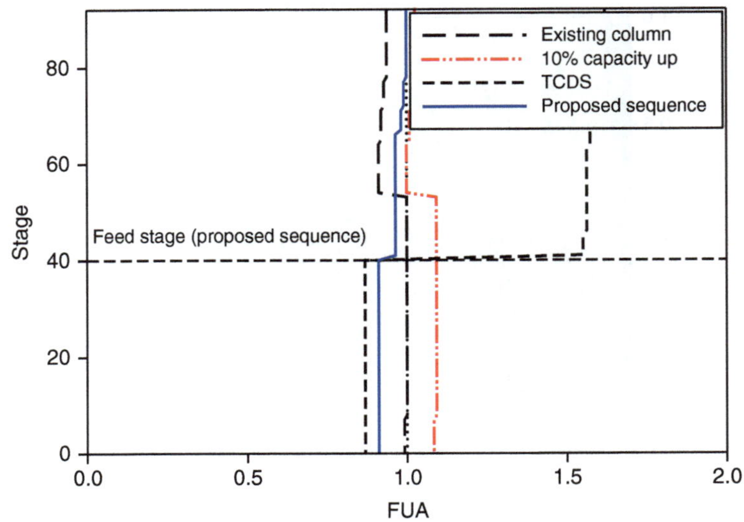

Fig. 4.53 FUA profiles of existing deisobutanizer (before and after capacity up); main column in TCDS; and main column in proposed sequence

the column or the use of a side heat pump reduces the vapor and liquid traffic above the tray, where the side heat pump is placed, and increases the vapor and liquid traffic below the tray. To solve this problem, the vapor and liquid traffic inside the column is redistributed using TCDS, and a side heat pump is utilized to move the vapor and liquid traffic from the flooding trays to the others, which can accommodate significant higher flows and reduce the energy requirements. In particular, when constructing the TCDS, the existing deisobutanizer was bottlenecked in the rectifying section, where FUA > 1, and less energy was required compared to the conventional column. The FUA curve identifies an opportunity to reduce the area utilization in the rectifying section of the column by increasing the flows in the stripping section of the main column. A side heat pump will condense the vapor phase and recycle to the column as a liquid phase. This causes a decrease in the vapor flow rate in the top section. No change in the top product was noted. Therefore, the reflux rate decreased from 123531 kg/hr to 82006 kg/hr compared to TCDS. Furthermore, the vapor in these trays increased due to an increase in the liquid of the trays below and there is no change in the bottom product. This caused an increase in vapor and liquid traffic. Accordingly, the use of a side heat pump can alter the vapor and liquid flows within a column section. In particular, the side heat pump can reduce the vapor and liquid traffic above the tray where it is replaced, and increase the vapor and liquid traffic below that tray. The hybrid technology combining a TCDS, which can save energy but cause flooding in the rectifying section, with a side heat pump, which can save energy, reduces the temperature difference compared to that using a heat pump in the top, and reduces the vapor and liquid traffic above the tray where the side heat pump is placed. In addition, it can save energy and remove the bottleneck problem efficiently.

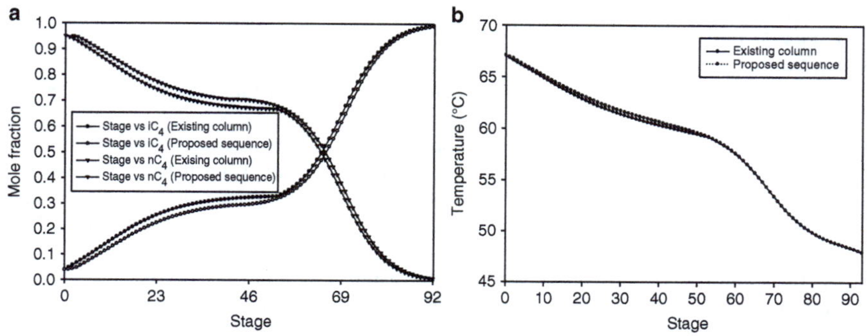

Fig. 4.54 (a) Composition profiles of iC_4 and nC_4 and (b) temperature profile of existing deisobutanizer and main column of proposed sequence

Figure 4.54a shows the composition profiles of iC_4 and nC_4 in the existing deisobutanizer and the main column in the proposed sequence. A portion of the vapor stream is condensed to the liquid phase, which causes an increase in the composition profile of the heavy component (nC_4) and a decrease in the composition profile of the light component (iC_4). As a result, the temperature profile was also increased compared to that in the existing deisobutanizer (Fig. 4.54b). This can decrease the temperature difference between intermediate vapor stream and bottom, which can reduce the side compressor duty.

4.10 Concluding Remarks

This chapter discussed several attractive techniques that can be applied to a multicolumn system in retrofit or debottlenecking. DWCs and TCDSs, which are growing steadily in the number and size of applications, are excellent choices for enhancing the performance of distillation sequences in retrofit and/or debottlenecking. Heat integration and process intensification or external and internal heat integration can simultaneously be employed to retrofit the conventional distillation sequences efficiently.

References

Agrawal R (1999) More operable fully thermally coupled distillation column configurations for multicomponent distillation. Chem Eng Res Des. 77:543–553

Agrawal R (2000) Multieffect distillation for thermally coupled configurations. AIChE J. 46: 2211–2224

Agrawal R, Fidkowski Z (1998) More operable arrangements of fully thermally coupled distillation columns. AIChE J. 44:2565–2568

Agrawal R, Fidkowski Z (1999) New thermally coupled schemes for ternary distillation. AIChE J. 45:485–496

Alatiqi IM, Luyben WL (1985) Alternative distillation configurations for separating ternary mixtures with small concentration of intermediate in the feed. Ind Eng Chem Process Des Dev. 24:500–506

Amminudin KA, Smith R (2001) Design and optimization of fully thermally coupled distillation columns. Part 2: application of dividing wall columns in retrofit. Trans IChemE 79:716–724

Amminudin KA, Smith R, Thong DYC et al (2001) Design and optimization of fully thermally coupled distillation columns: Part 1: preliminary design and optimization methodology. Trans IChemE. 79:701–715

Aspen Technology (2009) Aspen HYSYS thermodynamics COM interface. Version V7.1

Asprion N, Kaibel G (2010) Dividing wall columns: fundamentals and recent advances. Chem Eng Process. 49:139–146

Bildea CS, Dimian AC (1999) Interaction between design and control of a heat-integrated distillation system with prefractionator. Trans.IChemE 77 (Part A) 597–608

Blancarte-Palacios JL, Bautista-Valadés MN, Hernández Castro S et al (2003) Energy-efficient designs of thermally coupled distillation sequences for four-component mixtures. Ind Eng Chem Res. 42:5157–5164

Box GEP, Behnken DW (1960) Some new three level designs for the study of quantitative variables. Technometrics 2:455–475

Box GEP, Wilson KB (1951) On the experimental attainment of optimum conditions. J R Stat Soc Ser B 13:1–45

Caballero JA, Grossmann IE (2014) Optimal synthesis of thermally coupled distillation sequences using a novel MILP approach. Comput Chem Eng. 61:118–135

Carlberg NA, Westerberg AW (1989a) Temperature-heat diagrams for complex columns. 3. Underwood's method for the petlyuk configuration. Ind Eng Chem Res 28:1386–1397

Carlberg NA, Westerberg AW (1989b) Temperature-heat diagrams for complex columns. 2. Underwood's method for sidestrippers and enrichers. Ind Eng Chem Res 28:1379–1386

Cheng HC, Luyben WL (1985) Heat-integrated distillation columns for ternary separations. Ind Eng Chem Process Des Dev. 24:707–713

Christiansen C, Skogestad S, Lien K (1997) Complex distillation arrangements: extending the petlyuk ideas. Comput Chem Eng. 21:S237–S242

Dejanović I, Matijašević Lj, Olujić Ž (2010) Dividing wall column-a breakthrough towards sustainable distilling. Chem Eng Process. 49:559–580

Diggelen RCv, Kiss AA, Heemink AW (2010) Comparison of control strategies for dividing-wall columns. Ind Eng Chem Res. 49:288–307

Douglas JM (1988) Conceptual design of chemical processes. McGraw-Hill, New York, p 163–215

Dunnebier G, Pantelides CC (1999) Optimal design of thermally coupled distillation columns. Ind Eng Chem Res. 38:162–176

Dwivedi D, Strandberg JP, Halvorsen IJ et al (2012) Active vapor split control for dividing-wall columns. Ind Eng Chem Res. 51:15176–15183

Emtir M, Rev E, Fonyo Z (2001) Rigorous simulation of energy integrated and thermally coupled distillation schemes for ternary mixture. Appl Therm Eng. 21:1299–1317

Engelien HK, Skogestad S (2005) Multi-effect distillation applied to an industrial case study. Chem Eng Process. 44:819–826

Errico M, Rong BG, Tola G et al (2008) Process intensification for the retrofit of a multicomponent distillation plant-an industrial case study. Ind Eng Chem Res. 47:1975–1980

Errico M, Tola G, Rong BG et al (2009) Energy saving and capital cost evaluation in distillation column sequences with a divided wall column. Chem Eng Res Des. 87:1649–1657

Fenske MR (1932) Fractionation of straight-run Pennsylvania gasoline. Ind Eng Chem. 24: 482–485

Fidkowski Z, Krolikowski L (1986) Thermally coupled system of distillation columns: optimization procedure. AIChE J. 32:537–546

Gadalla MA, Olujic Z, Jansens PJ et al (2005) Reducing CO emissions and energy consumption of heat-integrated distillation systems. Environ Sci Technol. 39:6860–6870

Ghadrdan M, Halvorsen IJ, Skogestad S (2011) Optimal operation of Kaibel distillation columns. Chem Eng Res Des 89:1382–1391

Ghadrdan M, Halvorsen IJ, Skogestad S (2013) Manipulation of vapour split in Kaibel distillation arrangements. Chem Eng Process. 72:10–23

Glinos K, Nikolaides I, Malone F (1986) New complex column arrangements for ideal distillation. Ind Eng Chem Proc Des Dev. 25:694–699

Halvorsen IJ, Skogestad S (2003) Minimum energy consumption in multicomponent distillation. 2: three-product Petlyuk arrangements. Ind Eng Chem Res. 42:605–615

Hernández JGS, Hernández S, Jiménez A (2005) Analysis of dynamic properties of alternative sequences to the petlyuk column. Comput Chem Eng. 29:1389–1399

Hernández S, Jiménez A (1996) Design of optimal thermally-coupled distillation systems using a dynamic model. Trans Inst Chem Eng. 74:357–362

Hernández S, Jiménez A (1999) Design of energy-efficient petlyuk systems. Comput Chem Eng. 23:1005–1010

Huaqiang G, Xiangwu C, Nan C et al (2014) Experimental study on vapour splitter in packed divided wall column. J Chem Technol Biotechnol. 91:449–455

Jiménez A, Ramírez N, Castro A et al (2003) Design and energy performance of alternative schemes to the Petlyuk distillation system. Chem Eng Res Des 81:518–524

Kang KJ, Harvianto GR, Lee MY (2017) Hydraulic driven active vapor distributor for enhancing operability of a dividing wall column. Ind Eng Chem Res. 56:6493–6498

King ST, Haas J (2013) Vapor and liquid flow control in a dividing wall fractional distillation column. U.S. Patent 8,562,792 B2, 22 October

Kiss AA, Bildea CS (2011) A Control perspective on process intensification in dividing wall columns. Chem Eng Process. 50:281–292

Kiss AA, Ignat RM (2012) Innovative single step bioethanol dehydration in an extractive dividing-wall column. Sep Purif Technol. 98:290–7

Kiss AA, Suszwalak DJPC (2012) Innovative dimethyl ether synthesis in a reactive dividing-wall column. Comput Chem Eng. 38:74–81

Klemola KT, Ilme JK (1996) Distillation efficiencies of an industrial-scale i-butane/n-butane fractionator. Ind Eng Chem Res. 35:4579–4586

Kolbe B, Wenzel S (2004) Novel distillation concepts using one-shell columns. Chem Eng Process. 43:339–346

Lee B, Kim G (2012) Liquid splitter. Korean Patent Pending, 10-2012-0030223

Lee B, Kim G, Lee M, et al (2010) Liquid splitter. Korean Patent Pending, 10-2010-0120467

Lee MY, Minh LQ, Long NVD et al (2016) Retrofit of side stream columns to dividing wall columns, with a case study of industrial application. In: Rangaiah GP (eds) Chemical process retrofitting and revamping techniques and applications. Wiley, West Sussex, pp 251–283

Lee SK, Shin JH, Lee JK (2011) Divided wall distillation column for producing high purity normal butanol, and normal butanol distillation method. U.S. Patent US2011/0303526, 15 December

Liu ZY, Jobson M (2004a) Retrofit design for increasing the processing capacity of distillation columns 1. A hydraulic performance indicator. Chem Eng Res Des. 82:3–9

Liu ZY, Jobson M (2004b) Retrofit design for increasing the processing capacity of distillation columns. 2. Proposing and evaluating design options. Chem Eng Res Des. 82:10–17

Long NVD, Lee MY (2011) Improved energy efficiency in debottlenecking using a fully thermally coupled distillation column. Asia-Pac J Chem Eng. 6:338–348

Long NVD, Lee MY (2012a) Improvement of the deethanizing and depropanizing fractionation steps in NGL recovery process using dividing wall column. J Chem Eng Japan. 4:285–294

Long NVD, Lee MY (2012b) Dividing wall column structure design using response surface methodology. Com Chem Eng. 37:119–124

Long NVD, Lee MY (2012c) Improvement of natural gas liquid recovery energy efficiency through thermally coupled distillation arrangements. Asia-Pac J Chem Eng. 7:71–77

Long NVD, Lee MY (2013a) Optimal retrofit design of extractive distillation to energy efficient thermally coupled distillation scheme. AIChE J 59:1175–1182
Long NVD, Lee MY (2013b) Optimal retrofit of a side stream column to a dividing wall column for energy efficiency maximization. Chem Eng Res Des. 91:2291–2298
Long NVD, Lee MY (2013c) Design and optimization of heat integrated dividing wall columns for improved debutanizing and deisobutanizing fractionation of NGL. Korean J Chem. Eng. 30:286–294
Long NVD, Lee MY (2014) Review of retrofitting distillation columns using thermally coupled distillation sequences and dividing wall columns to improve energy efficiency. J Chem Eng Japan 47:87–108
Long NVD, Lee MY (2015) A hybrid technology combining heat pump and thermally coupled distillation sequence for retrofit and debottlenecking. Energy 81:103–110
Long NVD, Lee SH, Lee MY (2010) Design and optimization of a dividing wall column for debottlenecking of the acetic acid purification. Chem Eng Process. 49:825–835
Long NVD, Minh LQ, Pham TN et al (2016) Novel retrofit designs using a modified coordinate descent methodology for improving energy efficiency of natural gas liquid fractionation process. J Nat Gas Sci Eng. 33:458–468
Luyben WL (2004) Design and control of distillation columns with intermediate reboiler. Ind Eng Chem Res. 43:8244–8250
Malinen I, Tanskanen J (2009) Thermally coupled side-column configurations enabling distillation boundary crossing. 1. An overview and a solving procedure. Ind Eng Chem Res. 48:6387–6404
Manley DB (1997) Deethanizer/Depropanizer sequences with thermal and thermo-mechanical coupling and component distribution. U.S. Patent No. 5,673,571, 7 October
Mannan S, Fakhru'l-Razi A, Alam MZ (2007) Optimization of process parameters for the bioconversion of activated sludge by penicillium corylophilum, using response surface methodology. J Environ Sci 19:23–28
Olujić Ž, Jödecke M, Shilkin A et al (2009) Equipment improvement trends in distillation. Chem Eng Process. 48:1089–1104
Poth N, Brusis D, Stichlmair J (2004) Minimaler energiebedarf von trennwandkolonnen. Chem Ing Tech. 76:1811–1814
Premkumar R, Rangaiah GP (2009) Retrofitting conventional column systems to dividing wall columns. Chem Eng Res Des. 87:47–60
Rangaiah GP, Ooi EL, Premkumar R (2009) A simplified procedure for quick design of dividing-wall columns for industrial applications. Chem Prod Process Model., 4 Article 7. doi:10.2202/1934-2659.1265
Reid JA (2000) Distributed distillation with heat integration. Heating the process and fluid flow, PTQ AUTUMN 85–95
Rong BG (2011) Synthesis of dividing-wall column (DWC) for multicomponent distillations – a systematic approach. Chem Eng Res Des. 89:1281–1294
Rong BG, Turunen I (2006) Process intensification for systematic synthesis of new distillation systems with less than N-1 columns. Comput Aided Chem Eng. 21:1009–1014
Schultz MA, Stewart DG, Harris JM et al (2002) Reduce costs with dividing-wall columns. Chem Eng Prog. 98:64–71
Shin J, Lee S, Lee J et al (2011) Manage risks with dividing-wall column installations. Hydrocarbon Process. 90:59–62
Shin J, Lee J, Lee S et al (2013) Enhance operation and reliability of dividing-wall columns. Hydrocarbon Process. 92:85–88
Slade B, Stober B, Simpson D (2006) Dividing wall column revamp optimizes mixed xylenes production, IChemE, Symposium Series No. 152
Smith R (2005) Chemical process design and integration. Wiley, West Sussex, p 212–232
Triantafyllou C, Smith R (1992) The design and optimization of fully thermally coupled distillation. Trans Inst Chem Eng. 70:118–132

Turton R, Bailie RC, Whiting WB, Shaeiwitz JA, Bhattacharyya D (2012) Analysis, synthesis and design of chemical processes, 4th edn. Prentice Hall, Upper Saddle River
Tututi-Avila S, Jimenez-Gutierrez A, Hahn J (2014) Control analysis of an extractive dividing-wall column used for ethanol dehydration. Chem Eng Process. 82:88–10
Underwood AJ (1948) Fractional distillation of multicomponent mixtures. Chem Eng Progress 44:603–614
Vazquez-Castillo JA, Venegas-Sánchez JA, Segovia-Hernández JG et al (2009) Design and optimization, using genetic algorithms, of intensified distillation systems for a class of quaternary mixtures. Comput Chem Eng 33(11):1841–1850
Wankat PC (2012) Separation process engineering: includes mass transfer analysis, 3rd edn. Prentice-Hall, Upper Saddle River
Wenzel S, Rohm H (2003) Design of complex distillation columns by overall-cost optimization. Chem Eng Technol. 27:484–490
Wright SJ (2015) Coordinate descent algorithms. Math Program. 151:3–34

Chapter 5
Promising Retrofit Technologies for Nonconventional Distillation Sequence

5.1 Introduction

To achieve small amount of product or to remove small amount of impurities or to separate azeotropic mixtures or mixtures with similar boiling points, several non-conventional distillation sequences, such as batch distillation, side stream column, pressure swing azeotropic distillation, extractive distillation, azeotropic distillation, reactive distillation, and hybrid systems are preferred over conventional distillation sequences. Although they can accomplish the separation tasks, enhancing their performance has attracted considerable attention. Process integration and process intensification and their combination are again considered to be main options for retrofit and/or debottleneck these sequences.

5.2 Batch Distillation

5.2.1 Introduction

Continuous distillation is a thermodynamically efficient method of producing large amounts of material with a constant composition, while batch distillation has several advantages when small amounts of material or varying product compositions are required (Wankat, 2012). Advances in batch distillation have been associated with its use to distill alcohol, pharmaceuticals, coal oil, petroleum oil, and fine chemicals. This process is versatile with a run that can last from a few hours to several days.

As shown in Fig. 5.1a, heat is added through the jacket of the reboiler and extracted in the condenser, while the column is considered adiabatic. During the entire process the temperature of the reboiler increases, while the temperature of the top vapor increases or it is relatively constant depending on the operational policy (constant reflux ratio or constant distillate composition) applied (Modla and Lang, 2013).

N.V.D. Long, M. Lee, *Advances in Distillation Retrofit*, DOI 10.1007/978-981-10-5901-8_5

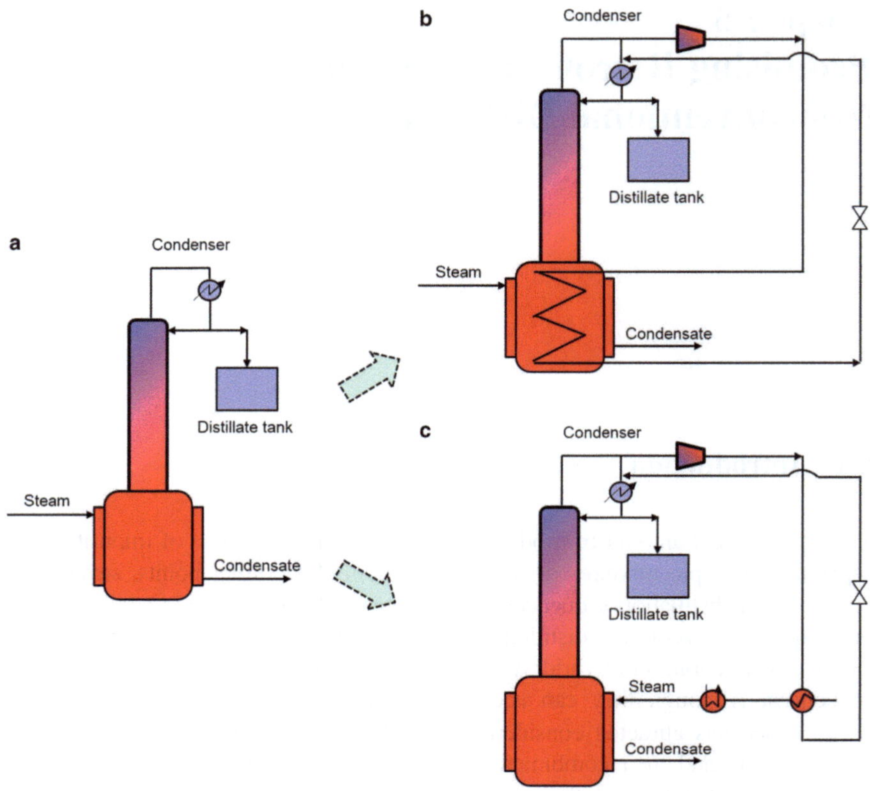

Fig. 5.1 Promising technology for the retrofit of batch distillation column using heat pump

5.2.2 *Retrofit of Batch Distillation*

The energy saving methods for batch distillation have attracted less attention compared to the continuous one. Heat pump systems are easy to introduce and the plant operation is generally simpler compared to other heat integration schemes, which are often more cost efficient owing to the simpler equipment (Harwardt and Marquardt, 2012; Jana, 2014). These appealing advantages have motivated researchers to explore the possibility of heat pump applications, particularly in batch processing. Basically, many heat pump concepts can be utilized. To have simple figures, the MVR heat pump was used to explore their applications in batch distillation. Figure 5.1b shows the promising technology for the retrofit of a batch distillation column using a heat pump. The top vapor is led to a compressor and then enters the inner coil of the reboiler, where it is condensed and cooled (the heat is transferred to the liquid in the reboiler). The stream condensed is partially refluxed to the top of the column after pressure reduction over a throttle valve. In this case, the compressor (the heat pump system) starts after the heating-up of the distillation column and works during the production step. Liquid expander can be utilized to replace the throttle valve and improve the energy efficiency of the proposed configuration.

Instead of transferring heat directly to the feed in reboiler, the top vapor stream can transfer heat to the usual heating medium (shown in Fig. 5.1c). In this system, there is no requirement for an inner heating coil in the reboiler and the compressor can be started as soon as the vapor appears at the top of the column. On the other hand, compared to the MVR system, at least one additional heat exchanger is needed, where the driving force of heat transfer (ΔT) must be ensured (Modla and Lang, 2013). In this case, the compressor (the heat pump system) can be already operated during the heating-up. Moreover, it is unnecessary to apply an additional inner coil for a reboiler. Note that a superheater should be located prior to the compressor to avoid the partial condensation of some saturated vapors, which can damage that compressor. This can enhance the performance of heat pump system with smaller compression ratio and power.

5.3 Side Stream Column

5.3.1 Introduction

Side stream columns (SSCs), in which the intermediate component is withdrawn by a side draw (Fig. 5.2), are considered to be more energetically favorable alternatives to utilize the remixing phenomenon, possibly leading to substantial energy consumption in the two-column sequences (Lee et al., 2016). An increasing number of SSCs are used in the chemical and petroleum industry because they offer a cost-effective way of producing three products from a single column in some separations (Papastathopoulou and Luyben, 1991). Basically, SSCs are convenient alternative schemes to sequence of simple columns when some limits of the product purity are acceptable or when those limits correspond to a very low impurity content of light and heavy boiling components (Niesbach et al., 2013; Brambilla, 2014). The main idea comes from the composition profiles of the first column of the two sequences. In the direct sequence, the composition of the intermediate component B in the first column increases below the feed as the more volatile component A decreases (Brambilla, 2014). On the other hand, moving further down the column, the less volatile component increases. Therefore, the composition of component B reaches a peak only to be remixed. At this highest point, one side stream is withdrawn from the vapor phase (Fig. 5.2a) (Brambilla, 2014). The second column of the sequence is replaced by only one stripping section with limited separation capability between component B and less volatile component C.

Similarly, composition B first increases above the feed as the less volatile component C decreases in the first column of the indirect sequence. The composition of component B reaches a maximum, then only to decrease with increasing more volatile component A. A liquid stream is suggested in the rectifying section to replace the second column in the indirect sequence with limited separation capability between component B and the more volatile component A (Fig. 5.2b).

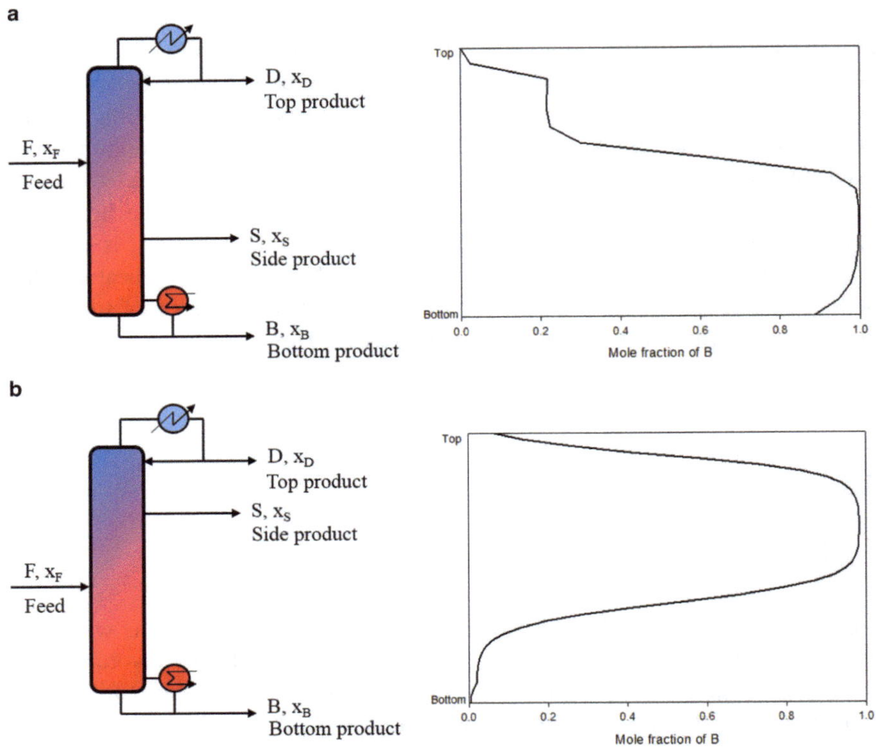

Fig. 5.2 Schematic diagram of a SSC (a) with a vapor side stream below the feed and corresponding composition profile of the intermediate component; (b) with a liquid side stream corresponding composition profile of the intermediate component

5.3.1.1 Heuristic Rules to Use SSCs

The SSCs can be used in some certain ranges of feed compositions and/or volatilities (Glinos and Malone, 1985). The following list of heuristics (Glinos and Malone, 1988) provides useful guidelines for SSC applications.

- SSC can be considered in the extent of the mole fraction of A and/or C < 0.2.
- Consider using a side stream when the column is used to recycle the reactants back to the reactor from a liquid separation system in a reaction-separation-recycle process.
- Consider a side stream above the feed when the intermediate component is more difficult to separate from the heavier components than from the light components. Otherwise, consider a side stream below the feed (note that heuristic 1 must be satisfied).

The purity of the side stream product is restricted by thermodynamics and the nature of the distillation process, which causes these columns be suitable as prefractionators, whose the side stream is fed to another column for further separation. In addition, they are used to generate recycle streams when there is no strict requirement on the recycle composition (Glinos and Malone, 1985). In addition, SSCs might be useful when the middle boiling components are trace components or even the main product, as in a pasteurization column (Rooks et al., 1996). The pasteurization column is a SSC used to remove the light-ends in the vapor phase using a full-reflux condenser.

In summary, in the ideal case, SSCs can be attractive either when the middle product dominates in the feed and one of the other components is found in only minor quantities, or when the middle boiling components are trace components or when they are used as prefractionators (Long et al., 2015). Therefore, SSCs can be applied only under special circumstances for the feed composition and the product specification (Smith, 2005).

5.3.1.2 Pros and Cons of a SSC

In order to enhance the energy performance of conventional configurations for ternary mixture separations, a SSC can be used (Glinos et al., 1986; Emtir and Etoumi, 2009). In particular, a side stream is introduced at the peak point composition of the middle components as a transfer stream to the second column to reduce the remixing effects in the first column. SSCs can also be employed in binary separation if different purity levels are desired (Papastathopoulou and Luyben, 1991). For example, a single column can be used to produce two grades of propylene products, where the bottom stream is propane, while the medium-purity propylene and high-purity polymer-grade propylene are collected from the side and distillate streams, respectively.

On the other hand, the common disadvantage of SSCs is the limitation of the purity of the intermediate component product stream (Nikolaides and Malone, 1987). This is because achieving a high-purity side stream may require high reflux ratios and a large number of stages, which also leads to higher purity in the distillate and bottom products (Glinos and Malone, 1985; Wachter et al., 1988). Requiring high reflux ratio and large number of stages can lead to a very high operating and investment cost.

5.3.2 Retrofit of a SSC to a DWC

Improving the energy efficiency of SSCs and reducing the related CO_2 emissions is a part of the global efforts towards greater sustainability in processing industries. Furthermore, increasing the capacity, which has been a major focus of the chemical process industry, can cause entrainment flooding or a bottleneck problem in the SSC. In general, a retrofit project in a distillation process covers a wide range of modifications and the utilization of existing processes. This may include only simple modifications or the replacement of column internals to large

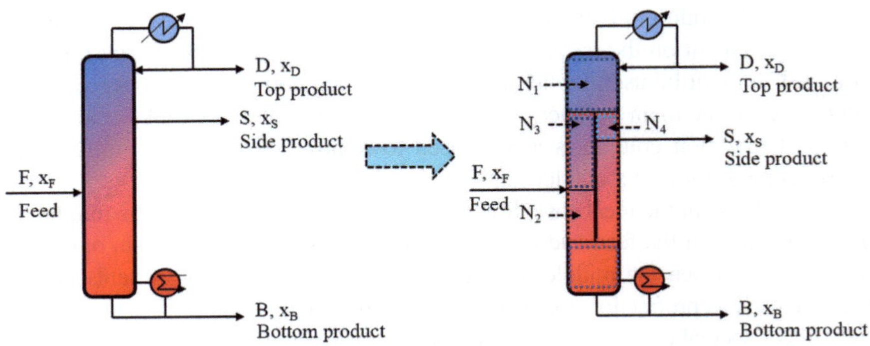

Fig. 5.3 Schematic diagram of a SSC and retrofitted DWC

modifications of the column configuration, partial enlargement of the shell diameter and height by replacing a part of the column or increasing the number of trays, and the modification of auxiliary equipment (Long et al., 2010). Amminudin and Smith (2001) claimed that the key to a successful retrofit lies in exploiting the existing hardware by maximizing the use of existing equipment while, at the same time, minimizing the need for new hardware to minimize the capital cost.

The aim of most retrofit projects for distillation columns is to reduce the energy requirements while maintaining the high purity and recovery of main product. As described above, SSCs consume a considerable amount of energy because of the inherent thermodynamic restrictions and the nature of the distillation process (Glinos and Malone, 1985). The retrofit of a SSC to a DWC (Fig. 5.3) can save significant energy (Lee et al., 2011; Long and Lee, 2013a). The successful exploitation of a DWC in retrofitting was also reported by ExxonMobil (Ennenbach et al., 2000; Kolbe and Wenzel, 2004; Slade et al., 2006).

5.3.3 *Retrofit of a SSC to a Heat Pump Assisted Dividing Wall Column (HPDWC)*

Figure 5.4 shows a number of options for retrofitting an existing SSC to a HPDWC. A heat pump is used to supply heat to the reboiler, the vessel of which is installed with a dividing wall (shown in Fig. 5.4a). The latent heat of condensation of the overhead can be recovered to provide partial/total heating of the reboiler. On the other hand, instead of supplying heat to the reboiler, a heat pump can be used to provide side heating (Fig. 5.4b and 5.4c), reduce the temperature difference and the traffic in the stripping section, and further increase the capacity. Similarly, a side heat pump assisted DWC is implemented to supply heat to the reboiler while reducing the traffic in the rectification section (Fig. 5.4d) or in the dividing wall

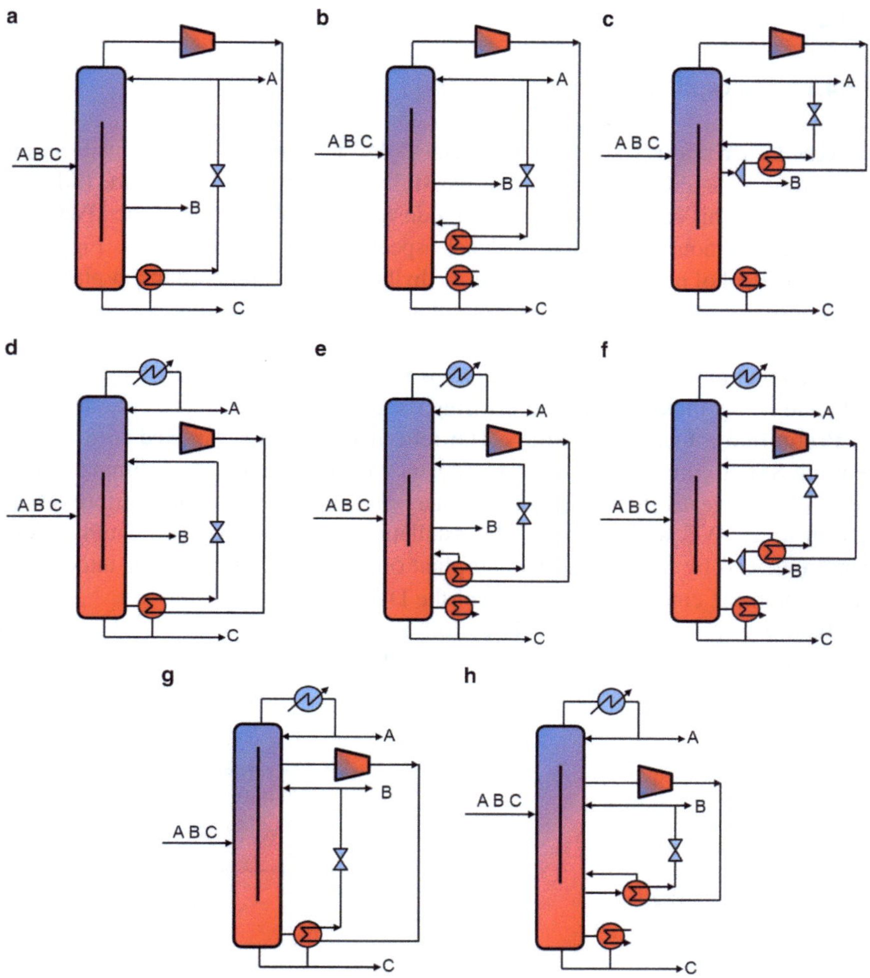

Fig. 5.4 Schematic diagram of the retrofitted HPDWCs

section (Fig. 5.4g). Supplying heat to the side reboiler by a side heat pump can reduce the temperature difference throughout the system and decrease the traffic in the affected sections (above the side heat pump and below the side reboiler, Fig. 5.4e, 5.4f and 5.4h). Consequently, this system can be applied not only for close-boiling, but also for wide-boiling mixtures. In addition, it can be employed both at the time of construction and as a retrofitted design. No change in diameter or the total number of stages of each column can be considered to highlight the optimal use of existing columns. The HPDWC was designed using a divided-wall and a heat pump. The required modifications for the retrofit may include a change in the feed tray and draw tray for the side stream product.

5.3.4 *Case Study*

5.3.4.1 Conceptual Design

Process Description The first commercial DWC in Korea was developed through the retrofit of a SSC and installed successfully in a plant of LG Chem. Ltd. (Lee et al., 2011). This case study reports the industrial implementation of the DWC with particular attention paid to the practical aspects of the retrofit project. Figure 5.5 shows an alcohol purification unit in a 2-ethylhexanol (2-EH) production plant. This unit consists of three main distillation columns: a prefractionation column, a 2-EH refining column, and a normal butyl alcohol (nBuOH) refining column. The nBuOH refining column is a SSC, where nBuOH is recovered from the side draw, while the light and heavy impurities are cut off from the top and bottom of the column.

Preliminary Feasibility Study The retrofit project was targeted to a nBuOH refining SSC to improve the energy efficiency of the column. A preliminary feasibility study revealed that considerable energy savings can be achieved through the retrofit to a DWC with a reasonable installation cost. The feasibility study results were reviewed by field engineers and plant operators to improve their understanding of the basic concept and benefits of the DWC.

Basic Design and Optimization Motivated by the promising results from the preliminary feasibility study, more rigorous studies were performed for the design and optimization of a retrofitted DWC. The DWC structure, such as the total

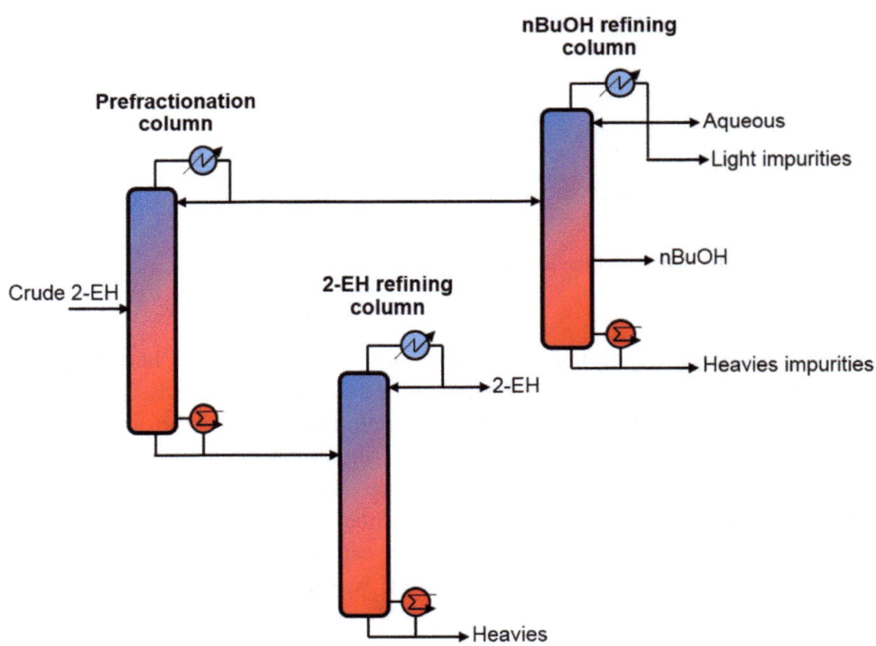

Fig. 5.5 Alcohol purification unit in a 2-EH production plant

number of theoretical trays, feed and product locations, was optimized rigorously while maintaining the same column height as the existing column. The liquid and vapor split ratios into the dividing wall section were also optimized. The construction costs were estimated considering the site-specific concerns. An internal change from existing random packing to advanced structured packing was considered to keep the same column height and improve the separation performance. Figure 5.6 shows the SSC and the retrofitted DWC configurations.

The Aspen Hysys™ was used to simulate the process in this study. The plant operation data was also used to ensure the reliability of the simulation results. The installation of a dividing wall does not always guarantee enhanced energy efficiency. The benefits are system specific and dependent on the properties of the separation system. The potential benefits and inherent problems were estimated thoroughly by carrying out an extensive simulation study.

At the top of the dividing wall section, the liquid coming down from the rectifying section can be split in a controlled manner between the two sides of the wall using a total liquid trap-out tray and sending a part of the total liquid to the prefractionator side and the remainder to the side stream side (Kiss and Bildea, 2011). An internal liquid splitter is available for adjusting the liquid flow split ratio. At the bottom of the divided wall section, the vapor is split in proportion to the cross-sectional area of each side and the hydrodynamic condition (i.e., pressure drop and flow resistance). The overall friction of each side is fixed by the internal geometry, which is already set at the design stage and cannot be changed later during operation. The vapor split is normally not adjustable during operation for control purposes. On the other hand, the internals can still be designed and implemented to give a desirable vapor split ratio by adjusting the friction of the bottom end section of the dividing wall. Therefore, extensive sensitivity studies were carried out to examine the possible variations of the actual vapor splitting ratio to assess the effects of unexpected deviations from its normal design conditions. The effects of the changes in the feed composition, liquid split ratio and other operating variables were modeled to ensure that the performance and product quality be maintained within the acceptable ranges. Note that a novel active vapor distributor was proposed recently to address the potential problems associated with a mechanically driven vapor splitter as discussed in Sect. 4.6.4.

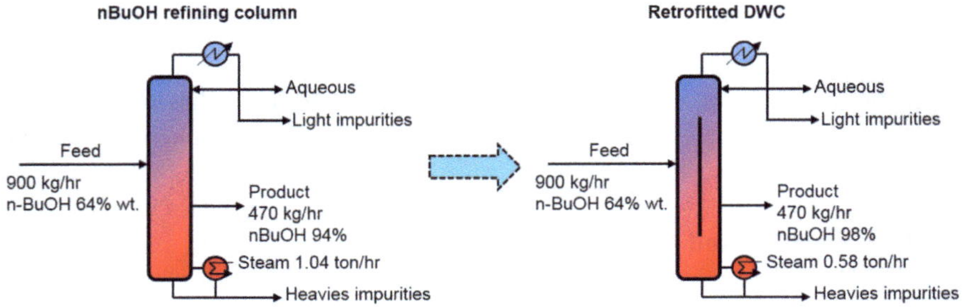

Fig. 5.6 nBuOH refining column: (a) before and (b) after retrofit to DWC

5.3.4.2 Detailed Design and Fabrication

According to the basic design study results, a detailed design was carried out including a fabrication plan for the column, internal, and several subsidiary devices. In the column internal design, special care was taken on the internals in the dividing wall section to ensure that the vapor split ratio was as designed. The pressure drops in the dividing wall section were predicted precisely and analyzed using the hydraulic correlations from the column internal suppliers and by rigorous computational fluid dynamics (CFD) analysis. A simple correlation was satisfactory for predicting the pressure drops with reasonable accuracy.

One important device in a DWC is an internal reflux splitter to manipulate the split of the liquid coming from the column overhead into the dividing wall section. A moving bucket type splitter was designed and fabricated to provide a precise split adjustment. Because of the potential risks from restless movement of its mechanical parts, it was replaced with a manual type reflux splitter in the next turnaround period.

The key to a successful retrofit lies normally in exploiting the existing equipment. To evaluate the possibility of reusing the original column, its flooding factor was checked to analyze the hydraulic performance. The diameter of the existing column was large enough to use it for a new DWC. On the other hand, a decision was made to fabricate a new one in advance to replace the existing column because of the tight turnaround schedule and the dominant labor cost. As the arrangements of nozzles and pipes are complicated and potentially confusing, a 3-D visual model was developed to help reduce human error in construction.

5.3.4.3 Foundation and Instrumentation

For a new column installation, sufficient structural strength is needed for the increased height and weight of the column after retrofitting. The retrofitted column consisted of five packing sections, which are the same as in the previous column. On the other hand, the different type of packing and additional internals installed in the DWC increased the total weight of the column. Therefore, the necessary civil work was performed to reinforce the foundation in the turnaround period (Fig. 5.7). The plant information system (PIS) and the screens of the distributed control system (DCS) were modified along with the additional instruments and control loops. The increased number of temperature sensors could help the engineers monitor the column operation and detect any abnormal behavior. The installation work was completed successfully within the scheduled maintenance period.

5.3.4.4 Start-Up and Commissioning

A considerably lower feed rate should be operated during start-up or shut down. DWCs normally use less energy, indicating lower vapor and liquid loadings than

Fig. 5.7 Foundation reinforcement for DWC installation (Reproduced with permission from Moonyong Lee, Le Quang Minh, Nguyen Van Duc Long, and Joonho Shin)

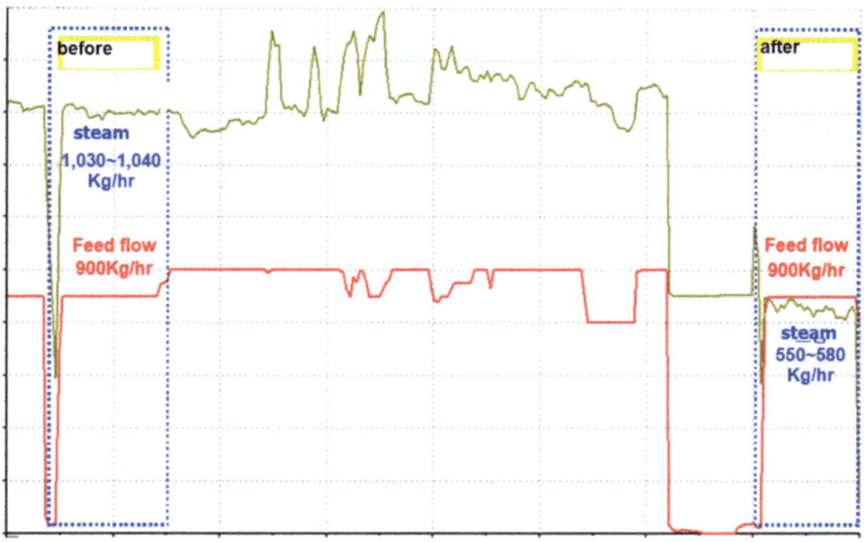

Fig. 5.8 Utility consumption before and after the retrofit (Reproduced with permission from Moonyong Lee, Le Quang Minh, Nguyen Van Duc Long, and Joonho Shin)

conventional columns, which might cause a turndown problem. The DWC was operated successfully at less than half of the normal load during start-up by increasing the boil-up and reflux rates up to a necessary level. The column temperature profile during normal operation matched the simulation results quite well. The performance of the DWC was successful. Steam consumption in the reboiler was reduced significantly (by 45%) after the retrofit to the DWC (Fig. 5.8). It is interesting to note that the nBuOH purity is increased dramatically from 94% to 98%, which brings an additional benefit.

From the preliminary study to actual implementation, the assistance of plant engineers and operators is essential for the success of the project. In many cases, process engineers who design DWC columns are unaware of the site specific concerns. Thorough and clear communication with them reduces the unnecessary time loss and risks during implementation. The concepts and operation modes of the DWC and new devices as well as the start-up and shutdown procedures were explained to the plant engineers and operators. Their questions were clarified using 3D models and simulation results.

5.4 Pressure Swing Azeotropic Distillation

5.4.1 Introduction

Azeotropic mixtures having vapor and liquid phases with identical compositions and mixtures with similar boiling points are commonly encountered in fine-chemical and specialty industries. Azeotropes occur because of non-ideal phase equilibrium resulting from the molecular interaction (either repulsion or attraction) of dissimilar chemical components (Luyben and Chien, 2010). The minimum-boiling azeotropes are caused by molecular repulsion between different types of chemical components, which can be either homogenous or heterogeneous, and are more common than systems in which molecular attractions occur that result in maximum-boiling azeotropes (Luyben, 2008a). When the repulsion is large enough, the repulsion forces can lead to the formation of heterogeneous minimum-boiling azeotropes (two liquid phases), which boil at a lower temperature than the boiling point of the lighter component (Luyben and Chien, 2010).

The separation of these mixtures is a challenging task in many chemical processes, and is impossible using a single conventional distillation column. Many non-conventional distillation techniques have been proposed to solve this problem (Widagdo and Seider, 1996; Doherty and Malone, 2001; Guedes et al., 2007; Gutiérrez-Guerra et al., 2009). The most common alternatives for homogenous binary azeotropic mixtures involve changing the operating pressure or adding a separation solvent. The sequences that involve changing the operating pressure, which are referred to pressure swing azeotropic distillation sequences, are employed when pressure has a strong effect on the azeotrope composition. Other alternatives include adding a light solvent referring to azeotropic distillation and adding a heavy extractive solvent referring to extractive distillation, which can cause a change in the relative volatility. The separation of heterogeneous binary azeotropic mixtures can be accomplished by taking advantage of the liquid-liquid phase separation in a decanter. A two-column system is used with the one liquid phase from a decanter fed to one column and the other liquid phase fed to the second column with high-purity products produced from the bottom of each column (Luyben and Chien, 2010). In this separation methodology, no third component is introduced into the system, which can avoid the many issues related to that component.

5.4.2 Pressure Swing Azeotropic Distillation

Pressure swings can be used in some systems for homogeneous binary azeotropes, whose azeotropic compositions change significantly with pressure (Luyben and Chien, 2010). As shown in Fig. 5.9, the composition of the azeotrope in this system changes according to the change in pressure. When this occurs, a two-column sequence operating at two different pressures can be used to achieve the desired separation, as shown in Fig. 5.10. The basic idea is to operate one column at low

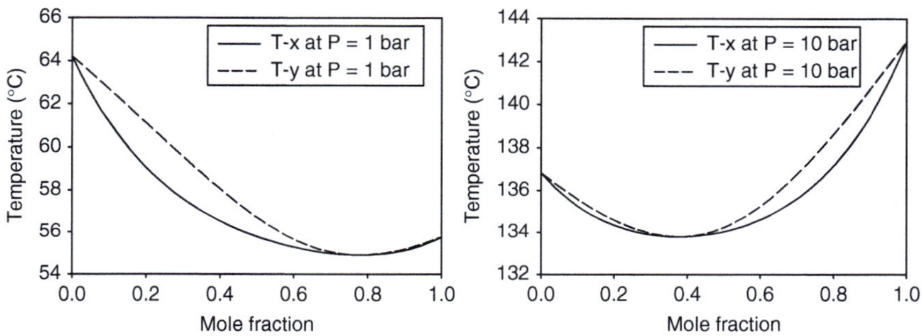

Fig. 5.9 Pressure sensitive, minimum boiling azeotrope

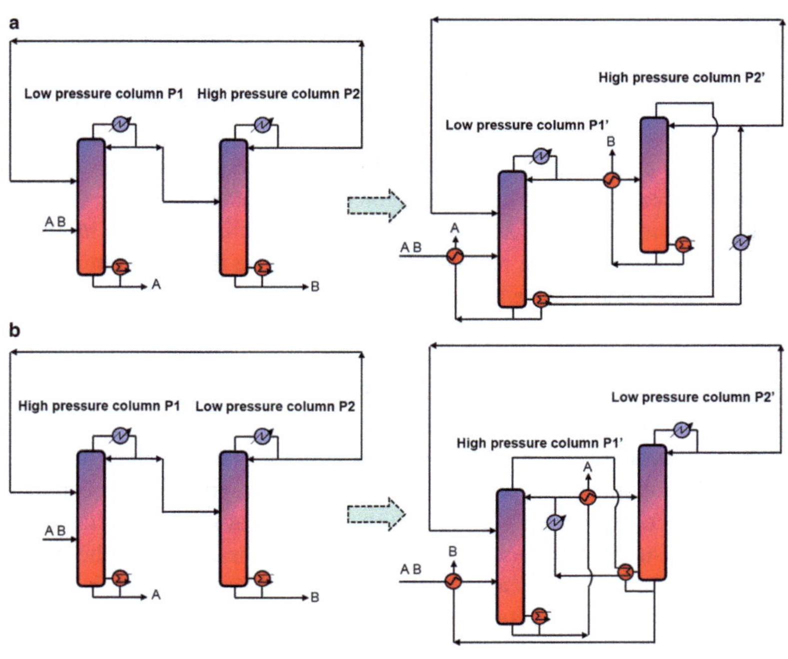

Fig. 5.10 Schematic diagram of pressure swing azeotropic distillation and the retrofitted pressure swing azeotropic distillation

pressure and another at high pressure. As shown in Fig. 5.10a, one of the components comes out from one end of the low-pressure column. If the azeotrope is minimum boiling, it will exit from the top and the product stream will exit from the bottom. The composition of this distillate stream will be close to that of the azeotrope at low pressure.

If the azeotrope is maximum boiling, the azeotrope and the product stream will exit at the bottom and top, respectively (Luyben and Chien, 2010). The azeotrope is then fed to the second high-pressure column, in which similar separation occurs. This azeotropic stream is fed back to the low-pressure column (Knapp and Doherty, 1992; Luyben, 2005; Luyben, 2008a; Luyben, 2008b; Wang et al., 2010; Yu et al., 2012). The large recycle streams are liquids that do not need to be vaporized. Therefore, if the azeotrope is maximum boiling, a pressure-swing system may be feasible despite showing relatively small changes in the azeotropic composition with pressure (Luyben and Chien, 2010). They claim that this inherent difference between minimum-boiling and maximum-boiling systems could be quite significant in terms of the flowsheet selection.

5.4.3 *Retrofit of Pressure Swing Sequence*

The operating pressure of each column and heat integration can be modified to increase the energy efficiency of pressure swing sequence. The important design variables are the operating pressures for each conventional distillation (Babi and Gani, 2014). The simple rule to select the operating pressure, P1, for the first column is to select the corresponding section of the azeotrope with the largest driving force, provided the feed mixture composition is also within the composition range of this azeotrope (Babi and Gani, 2014). In addition, the operating pressure for P2 corresponds to the separation with the driving force for the other region of the azeotrope but at the pressure, P2. Consider that reducing the operating pressure in a low pressure column can decrease the reboiler duty and increase the temperature difference between its base and the reflux-drum temperature in a high pressure column. Therefore, the objective is to stay as close as possible to the atmospheric pressure to reduce the operating costs and to select two operating pressures that give the largest driving forces for the two regions of the azeotropes and high temperature difference.

After modifying the operating pressure, heat integration can be considered. The reflux-drum temperature of the high-pressure column is higher than the base temperature of the low-pressure column. Therefore, heat integration can be attractive in terms of energy consumption (Luyben, 2008c). The heat released in the condenser of the high-pressure column as a portion or all of the heat input to the base of the low-pressure column. In the case a portion of the heat is available, the remaining heat can be provided by an auxiliary reboiler, which is normally heated by low-pressure steam. The reboiler in the high-pressure column might be driven by high-pressure steam. In the case where the two columns are fully heat integrated and no auxiliary reboiler is used (Fig. 5.10), the only energy input is the heat-transfer duty in the high-pressure column. Normally, the temperature difference between the base of the high pressure column and the reflux-drum temperature in high pressure column is high enough for heat integration. At this stage, the operating pressure can be modified in case the temperature is insufficient. Furthermore, the heat from the bottom streams of two columns and top vapor stream of the low pressure column can be utilized to preheat the feeds for further improving energy efficiency.

Modifying the operating pressure and heat integration are promising techniques to retrofit the pressure swing azeotropic distillation. Because a temperature difference exists in pressure swings azotropic distillation, heat integration is favorable and desirable.

5.5 Extractive Distillation Sequence

5.5.1 Introduction

To separate the azeotropic mixtures effectively, while the pressure option is economically feasible only for mixtures sufficiently sensitive to pressure, another alternative is the typical nonconventional distillation process encountered in industry and involves what is known as extractive distillation, which uses a heavy solvent that is generally nonvolatile, has a high boiling point and is miscible with the mixture. This solvent interacts differently with the components of the mixture, thereby causing a change in their relative volatility (Bravo-Bravo et al., 2010). This means that new three-part mixtures can be then separated by normal distillation. Moreover, the process allows a pure compound to be obtained at the top of the column with the other, together with the solvent, collected at the bottom, which can be separated easily in a secondary distillation or solvent recovery column due to the high boiling point of the solvent (Gutiérrez-Guerra et al., 2009).

Figure 5.11 shows the conventional extractive distillation sequence including two columns. The two-feed extractive column had an extractive column followed by a single-feed solvent recovery column. The upper section of the extractive column (above the extractive solvent feed location) is called the rectifying section separating component A and the extractive solvent. The middle section of the column, which is between the extractive solvent feed stage and the feed stage, is referred to as the extractive section preventing the ascent of component B in the column. The bottom section, which is below that feed tray location, is considered to be stripping section preventing descending of component A. The bottom stream of extractive distillation involving the component B and extractive solvent is then fed to the solvent recovery column to purify them so that the extractive solvent can be recycled back to the extractive column. A cooler is employed because the extractive solvent temperature can be considered as a design variable.

5.5.1.1 Criteria to Choose Extractive Solvent

Note that different extractive solvents might make the relative volatility between the components A and B to be greater or less than 1. In other words, the

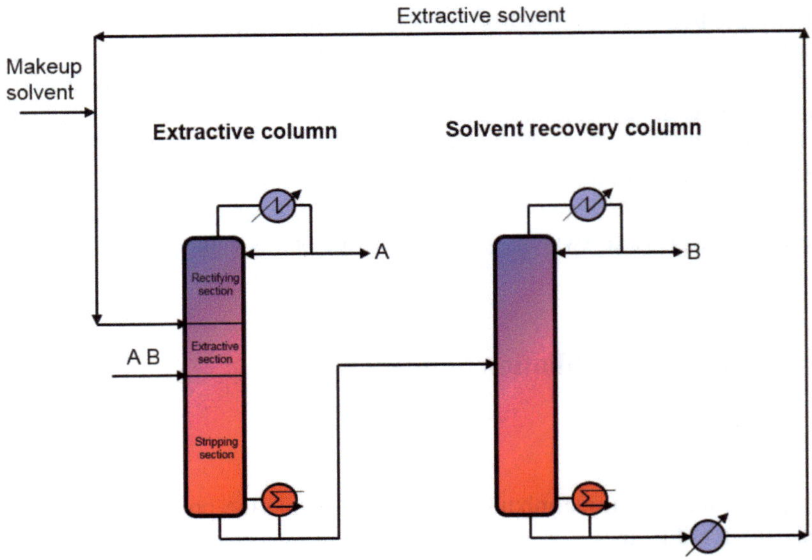

Fig. 5.11 Schematic diagram of a conventional extractive distillation sequence

performance of extractive distillation depends on the nature of the extractive solvent and component A can become a light or heavy component. For example, when water is chosen as the extractive solvent to separate acetone and methanol mixtures, the relative volatility between acetone and methanol is greater than one, that is acetone goes to the top of extractive distillation, while methanol becomes the light component and goes to the top when chlorobenzene is added to the extractive distillation (Luyben and Chien, 2010). Because extractive solvent selection plays an important role in the design of extractive distillation, many books and papers have analyzed and evaluated this issue. Normally, the selection of the solvent is chosen based on a set of essential, desirable, and environmental-health-safety (ESH) properties (Babi and Gani, 2014; Harper et al., 1999).

- Essential properties:
  - Be a liquid at the feed condition
  - Selectively dissolve one of the compounds in the mixture
  - Not form azeotropes with compounds A or B
  - Not react with any of the compounds
  - Have a sufficient difference in boiling point and vapor pressure with respect to compound A (so it can be recovered and recycled easily)

- Desirable properties:
  - Higher selectivity for A than B (so solvent loss would be low)
  - Higher solubility of A (compared to other solvents)
  - Low heat of vaporization (so the cost of the reboiler heat supply will be low)

- ESH properties:
  - Non-toxic
  - Not contributing to global warming potential
  - Not contributing to ozone depletion
  - Obtained from renewable sources

According to Hostrup et al. (1999), the Integrated Computer Aided System (ICAS) has a special tool for the selection of solvents for the separation of azotropes using extractive distillation. Luyben and Chien (2010) generated isovolatility curves, which is the curve that traces the unity-relative volatility of a binary mixture by adding an extractive solvent to extractive distillation, using Aspen Plus to select a suitable extractive solvent. According to the authors, the closer the intersection point is to the original component corner, the less extractive solvent is required, which means lower operating and capital costs. They also claimed that the equivolatility curves should be checked to determine how an extractive solvent enhances the relative volatility to choose a more effective extractive solvent.

5.5.2 *Retrofit Design and Optimization*

5.5.2.1 Design

Extractive distillation is used widely in a number of processes but its high energy consumption is the major main disadvantage, which requires dispatching. Process integration and process intensification are interesting options for solving this problem and they have been successful in reducing the energy requirements compared to processes where all the units are configured with little or no integration (Bravo-Bravo et al., 2010). In particular, the use of columns with thermal coupling has attracted considerable attention, providing potential cost savings (Fidkowski and Krolikowski, 1986; Fidkowski and Krolikowski, 1987; Agrawal and Fidkowski, 1999). Thermally coupled extractive distillation sequences with side rectifier (TCEDS-SR) obtained through the implementation of interconnecting streams can have lower energy consumption than conventional extractive distillation sequences (Shizou et al., 2000; Gutiérrez-Guerra et al., 2009; Ibarra-Sánchez and Segovia-Hernández, 2010; Long and Lee, 2013b).

Figure 5.12a presents the conceptual possibility of re-arranging two existing columns into a TCEDS-SR, where the extractive and side rectifier columns of the retrofit scheme utilize extractive and solvent recovery columns of the conventional scheme, respectively. Retrofitting the extractive distillation sequence begins with the development of preliminary designs for complex systems and minimizes the heat duty supplied to reboilers through optimization procedures. The TCEDS-SR column was designed using a thermal link in the vapor phase in the conventional direct sequence, which eliminates the reboiler in the second column of the conventional scheme (Gutiérrez-Guerra et al., 2009). Modification for the retrofit involves a change in the draw tray for the vapor and feed tray for liquid stream. More pipe

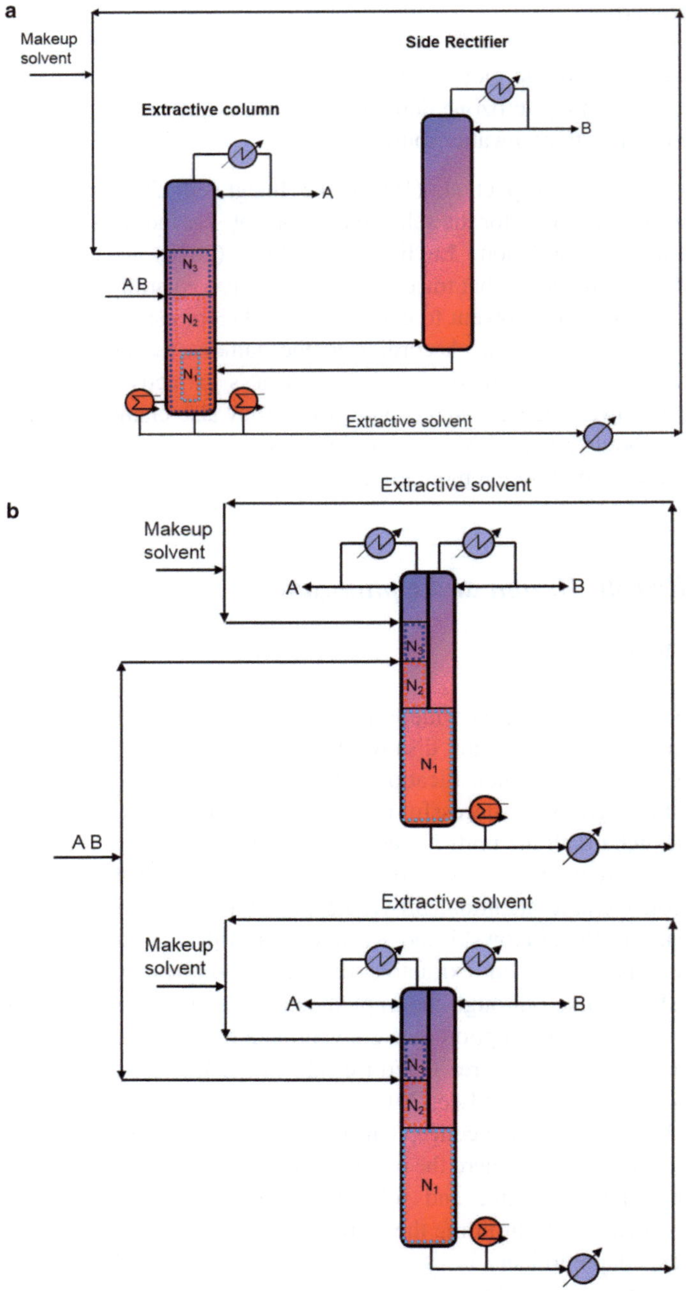

Fig. 5.12 Schematic diagram of (a) retrofitted TCEDS-SR and (b) retrofitted 2-parallel ETDWC sequence

work is also needed for the thermal link and re-assignment of the recovery column reboiler to the main column. Note that the bottom of retrofitted extractive column might have a bottleneck problem, which can be removed by adding a side reboiler.

Furthermore, the extractive top dividing wall column (ETDWC) could make the separation of an azeotropic mixture feasible and save considerable energy compared to the conventional extractive two-column sequence (Long and Lee, 2011). This configuration also offers a great chance for replacement by reducing the investment and space. Therefore, it could be considered in a retrofit of an extractive distillation sequence to reduce the energy requirements or increase the capacity. Figure 5.12b shows one promising configuration retrofitted from an existing conventional extractive distillation sequence by adding a dividing wall in the top of each column and one more condenser.

5.5.2.2 Optimization

To minimize the modifications in a retrofit, the number of trays of all columns was fixed to the same number in conventional extractive distillation columns. The location of the feed trays and the solvent/feed ratio have a substantial impact on energy requirements and products obtained. Here, the solvent/feed ratio was also kept constant as the optimal value in the conventional extractive distillation sequence. Therefore, after a preliminary design for the TCEDS-SR column and ETDWC, the main design variables, such as the internal vapor flow (FV) in the ETDWC or vapor flow (FV) to the side rectifier in the TCEDS-SR, vapor (N1), feed (N2), and solvent (N3) stream locations, were then optimized. The response surface methodology was used to examine the effects on the reboiler. After determining the preliminary ranges of the variables through single-factor testing, a Box-Behnken design was used to determine how the variables interact and to optimize the system in terms of the reboiler duty.

5.5.3 Case Study

5.5.3.1 Existing Plant Subject to Retrofit Study

Acetone and methanol are popular solvents and reagents in organic chemistry. Acetone is used extensively in industry to dissolve plastics, as a drying agent, and as a primary component in nail polish remover. Methanol is also a commonly used solvent, and its mixtures with acetone are normally observed, being necessary to separate volatile solvents for future applications (Gil et al., 2009). The acetone–methanol system has a minimum boiling point azeotrope, and extractive distillation is a possible method used to separate this azeotropic mixture, in which a heavy component is added to the mixture to increase the relative volatility of the original components. Extractive solvent selection is an important step because the separation effectiveness depends largely on the interactions between this component and the azeotropic mixture (Treybal, 1955; Kotai et al., 2007; Gil et al., 2009). Many studies have examined

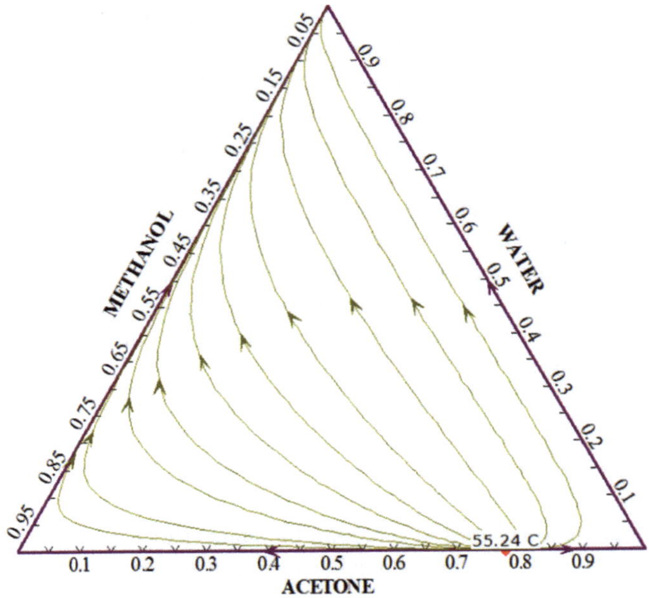

Fig. 5.13 Residue curve map of mixture acetone - methanol with water as an extractive solvent

the vapor–liquid equilibrium of this system and suggested water as a solvent (Verhoeve and Schepper, 1973; Iglesias et al., 1999; Vercher et al., 2006).

As shown in Fig. 5.13, a homogeneous minimum-boiling azeotrope exists at 1.01 bar with a composition of 77.6 mole% acetone at 55.24°C. In this sequence, there is no distillation boundary that divides the composition profile into different regions, and separation was feasible. Figure. 5.14 presents the process flow diagram of the existing extractive distillation process. The process has two columns, one for extractive separation and the other for solvent recuperation. The optimal design had a 57-stage extractive column and a 26-stage methanol column, both operating at 1.01 bar (Luyben and Chien, 2010). Note that the solvent is fed to the extractive column on a tray above the feed tray and below the top of the column so that high-purity acetone can be produced in the distillate (Luyben, 2008b). The UNIQUAC model was used to describe the thermodynamic properties. The solvent flow rate must be 1210 kmol/h to achieve the desired 99.5 mole% acetone purity.

5.5.3.2 Retrofitted TCEDS-SR or retrofitted ETDWC

For showing the benefits when retrofitting the existing process, TCEDS-SR was considered. Note that the 2-parallel ETDWC sequence can also be an attractive option. The TCEDS-SR column was designed using a thermal link in the vapor phase in the conventional direct sequence, which eliminates the reboiler in the second column of the conventional scheme. Modification for the retrofit involves a change in the draw trays for the vapor and liquid stream. More pipe work is also needed for the thermal

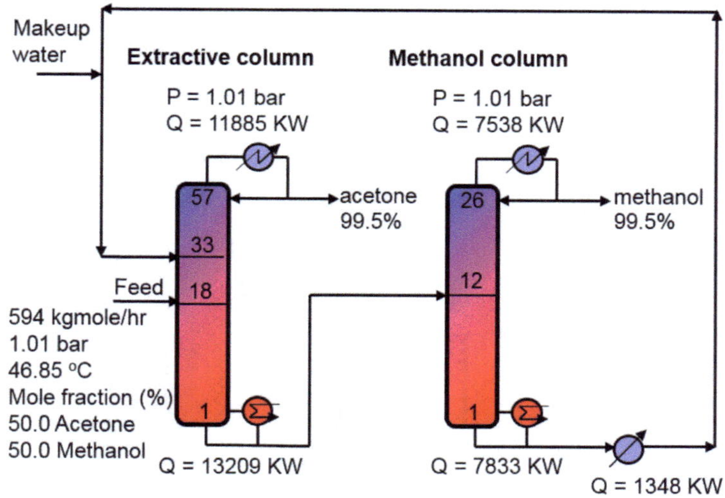

Fig. 5.14 Simplified flow sheet illustrating the existing extractive distillation sequence

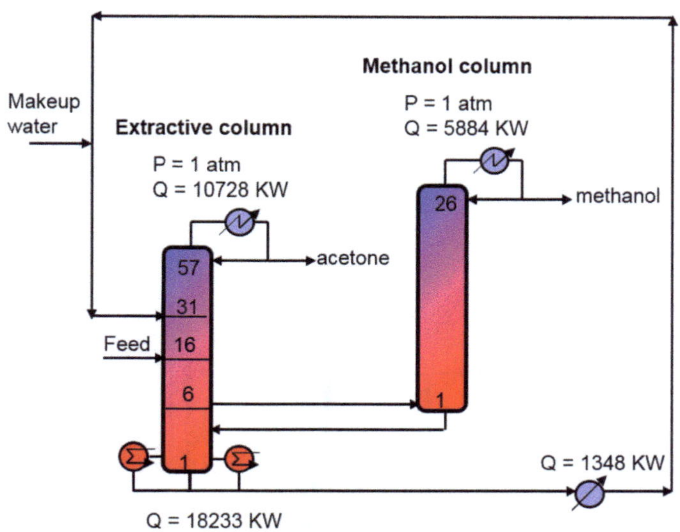

Fig. 5.15 Simplified flow sheet illustrating the retrofitted TCEDS-SR

link and re-assignment of the recovery column reboiler to the main column. Figure 5.15 presents a simplified flow sheet illustrating the retrofitted TCEDS-SR for the separation of acetone-methanol mixtures. After the optimization procedure, the simulation showed that the TCEDS-SR consumes 18233 KW, which is in good agreement with the results predicted by the RSM (18144 KW). Considerable savings (13.3%) in reboiler energy could be achieved. The reboiler in the methanol column can be used together with the existing reboiler in the extractive column.

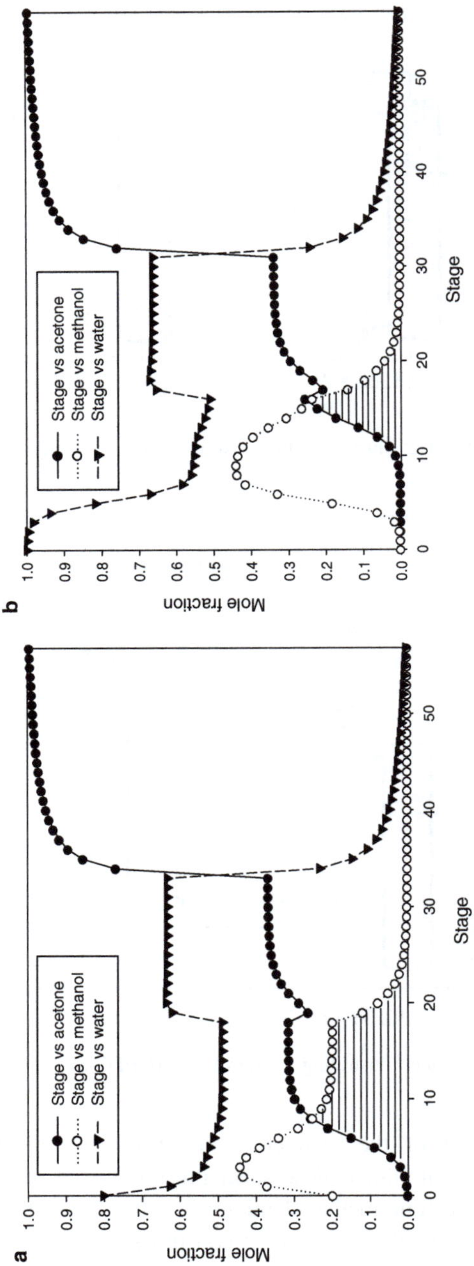

Fig. 5.16 Composition profiles for the extractive column of (a) conventional extractive distillation sequence and (b) retrofitted TCEDS-SR

Figure 5.16 presents the composition profiles of the extractive column for the conventional extractive distillation sequence and the TCEDS-SR sequence. The composition profile of methanol was quite smooth with no pinch zones in the retrofitted TCEDS-SR. The separation zone of acetone and methanol was lower than that in the conventional extractive distillation sequence. Figure 5.17 presents the composition profiles for the solvent recovery column of the conventional extractive distillation sequence and the side rectifier of the retroffited TCEDS-SR. Pinch zones also appeared from the 9th stage to the 12th stage for both the acetone and methanol composition profiles, which were eliminated in the retrofitted TCEDS-SR. Furthermore, the maximum flooding in the retrofitted TCEDS-SR was reduced by 79% compared to the conventional extractive distillation sequence (85%), as shown in Fig. 5.18, which offers 8% additional capacity.

Vapor is normally transported when a compressor applies a pressure difference between the inlet and outlet, which results in more investment and operating cost. To eliminate the use of a compressor, the extractive distillation can be designed with higher operating pressure so that the pressure of the vapor stream tray location is higher than that of the bottom section of the side rectifier.

5.6 Azeotropic Distillation Sequence

5.6.1 Introduction

Instead of adding a heavy solvent, a light entrainer is added to alter the relative volatility of one of two original components more than the other so that it can carry one of the components overhead in a distillation column. The two components to be separated are often close boiling point components that do or do not have an azeotrope point on the binary mixture, but sometimes they are components that form an azeotrope, even though they are not close boiling point components (Lei et al., 2005). Normally, azeotropic distillation has two types: homogeneous azeotropic distillation and homogeneous azeotropic distillation. In homogeneous azeotropic distillation, a phase split does not appear in the liquid along the entire column (Fig. 5.19a), while two liquid phases exist in some regions of the composition space in heterogeneous azotropic distillation.

Heterogeneous azeotropic distillation is used more widely for separating the close boiling point components or azeotropes than homogeneous azeotropic distillation (Lei et al., 2005). Figures 5.20a, 5.21a, and 5.22a illustrate three possibilities of heterogeneous azeotropic distillation. As shown in Fig. 5.20a, the sequence involves a preconcentrator column and heterogeneous azeotropic distillation. One example is the separation of pyridine – water mixtures having an azeotrope with composition of approximately 77 mole% water by adding toluene as an entrainer (Luyben and Chien, 2010). In the separation sequence shown in Fig. 5.20a, the purpose of the preconcentrator is to remove some of the water in the bottom stream, while the distillate composition approaches the azeotropic

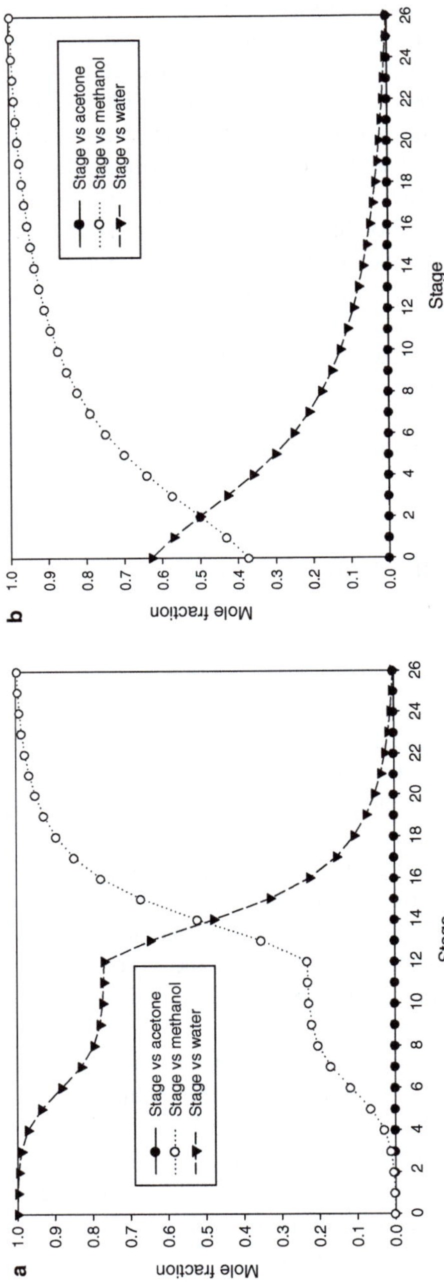

Fig. 5.17 Composition profiles for the (a) solvent recovery column of the conventional extractive distillation sequence and (b) side rectifier of the retrofitted TCEDS-SR

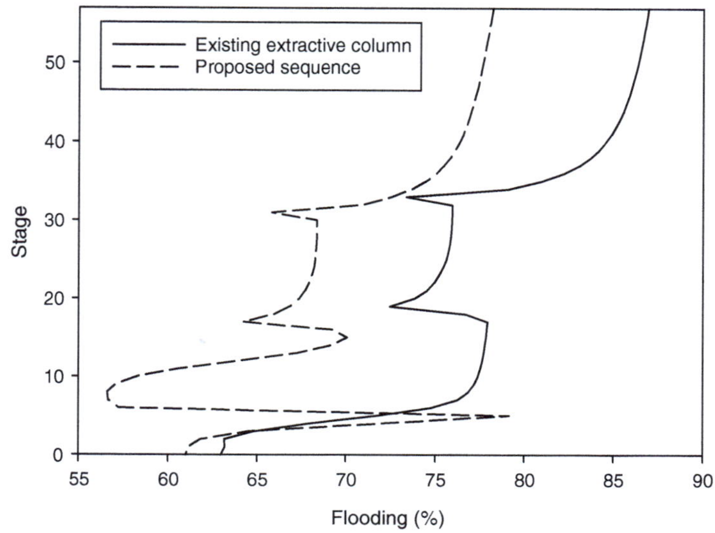

Fig. 5.18 Flooding profiles for the extractive column of existing extractive distillation sequence and retrofitted TCEDS-SR

composition of pyridine-water. That distillate stream is then fed into a heterogeneous azeotropic column. With the aid of adding another component as an entrainer (toluene), one desirable heterogeneous azeotrope (toluene–water), which has lowest temperature, is formed. Therefore, the heterogeneous azeotropic column top vapor is designed to approach a toluene–water azeotrope and the bottom product is pure pyridine. By condensing and cooling the top vapor and then sending it to a decanter, natural liquid-liquid separation into an organic phase and an aqueous phase will occur. The organic phase, which is rich in entrainer, is designed to return to the heterogeneous azeotropic distillation column. The composition of the aqueous phase is pure enough that it can be drawn out of the system as an aqueous product. A small entrainer makeup stream is required to feed into the system at the decanter to balance the small entrainer loss in the aqueous stream. To the system, such as an isopropanol-water system with cyclohexane as the entrainer, the aqueous phase still contains a significant amount of the main product; it is needed to feed that aqueous phase into a recovery column (Fig. 5.21a) (Luyben and Chien, 2010). In this recovery column, pure water can be obtained at the bottom, while the distillate is recycled back to the heterogeneous azeotropic column. Another design 3 was also proposed to improve designs 1 and 2 of heterogeneous azeotropic distillation sequence (Fig. 5.22a). The top vapor of the heterogeneous azeotropic column does not need to be placed strictly at the heterogeneous azeotrope. The second column in this sequence serves as a preconcentrator column and also as a recovery column to obtain a pure water product. More details of the design of these sequences can be referred to book by Luyben and Chien (2010). Here, promising configurations for retrofit these sequences to improve energy efficiency and/or increase capacity are proposed.

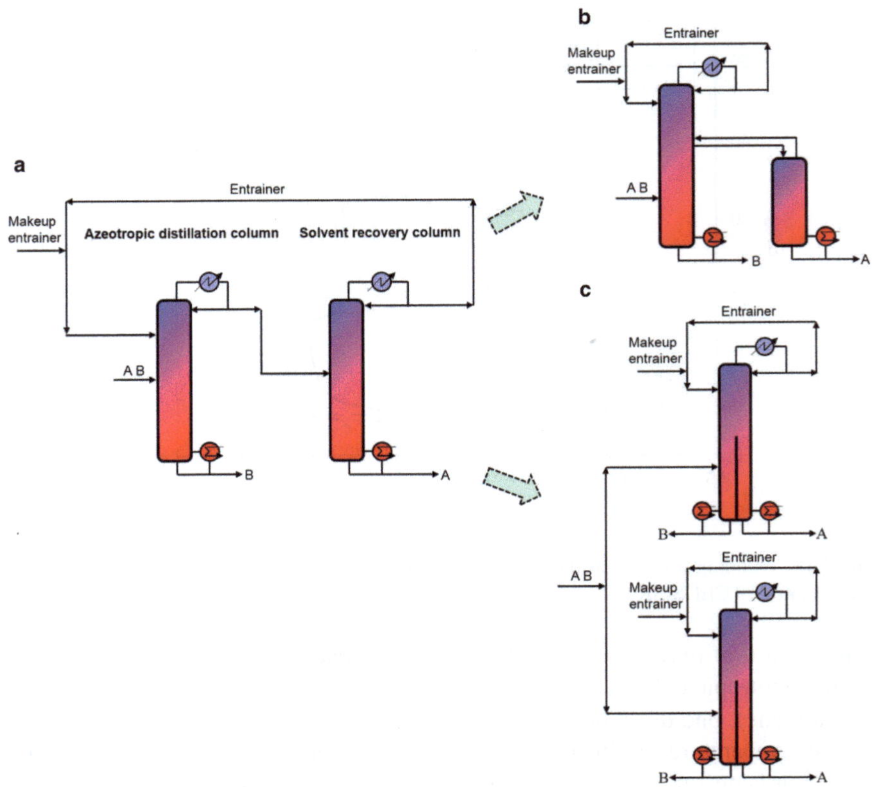

Fig. 5.19 Schematic diagram of (a) homogenous azeotropic distillation sequence; (b) retrofitted thermally coupled homogeneous azeotropic distillation sequence and (c) retrofitted 2-parallel ABDWC sequence

5.6.2 *Retrofit of Azeotropic Distillation*

Figures 5.19b and 5.19c shows two conceptual possibilities of retrofitting two existing columns of a homogeneous azeotropic distillation sequence into a thermally coupled azeotropic distillation scheme and a 2-parallel azeotropic bottom dividing wall column (ABDWC) sequence, respectively. To minimize the reboiler duty, the main design variables, such as the internal liquid flow (FL) in the retrofitted ABDWC or liquid flow (FL) to the side stripper in the retrofitted thermally coupled azeotropic distillation sequence, vapor, feed, and solvent stream locations, are optimized. Similarly, Fig. 5.20b and 5.20c shows two promising configurations to improve design 1 of the heterogeneous azeotropic distillation sequence (Long and Lee, 2014). A multi-effect technology can be considered to improve the energy efficiency for designs 2 and 3 of the heterogeneous azeotropic distillation sequence (Fig. 5.21b and 5.22b). The performance of designs 2 and 3 can also be enhanced by heat pump technology. By installing a compressor or a

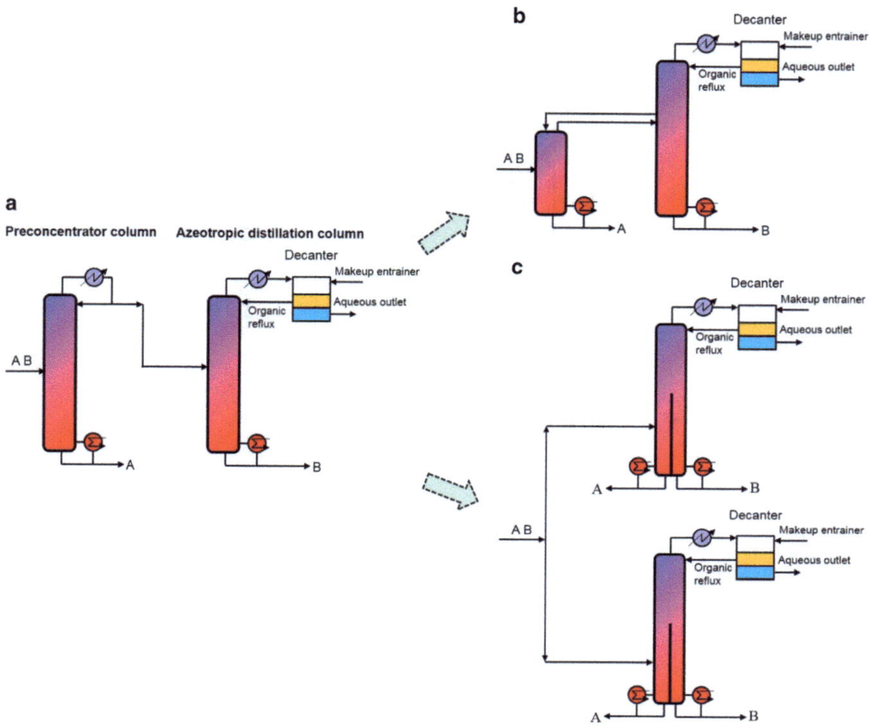

Fig. 5.20 Schematic diagram of (a) design 1 of heterogeneous azeotropic distillation sequence; (b) retrofitted thermally coupled heterogeneous azeotropic distillation sequence and (c) retrofitted 2-parallel ABDWC sequence

blower in the top of recovery column in design 2 and preconcentrator/recovery column in design 3, the latent heat of condensation of these columns overhead can be recovered to provide a partial/total heating duty at the reboiler of the azeotropic distillation column.

5.7 Reactive Distillation

5.7.1 Introduction

Reactive distillation is one of the most attractive proven configurations of process intensification, in which the reaction and distillation occur at the same time and in the same location. Compared to conventional reaction-distillation sequences, the so-called reactive distillation processes enable higher reactant conversion, product selectivity, as well as lower energy, water, and solvent consumption, leading to reduced investment and operating costs (Keller, 2014). Some of these benefits are realized by using the reaction to improve separation (e.g., overcoming azeotropes,

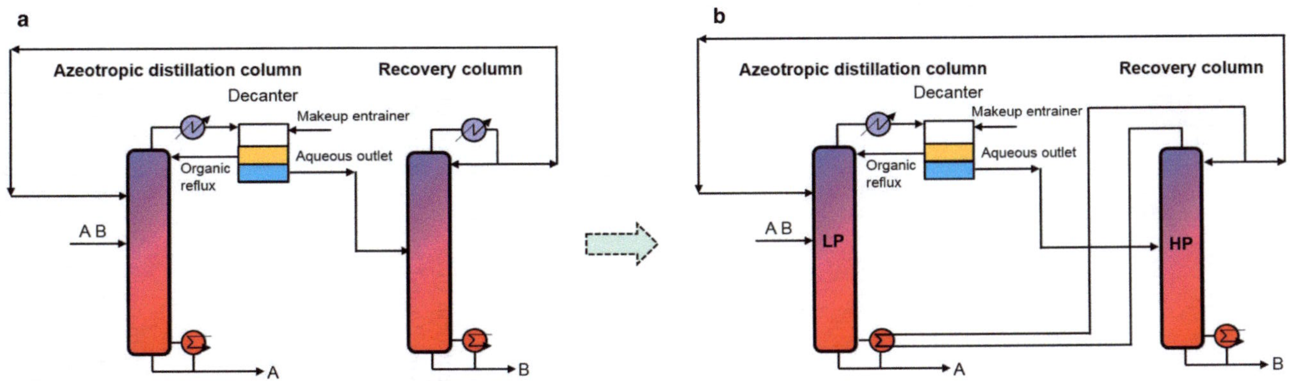

Fig. 5.21 Schematic diagram of (a) design 2 of heterogeneous azeotropic distillation sequence and (b) multi-effect heterogeneous azeotropic distillation sequence

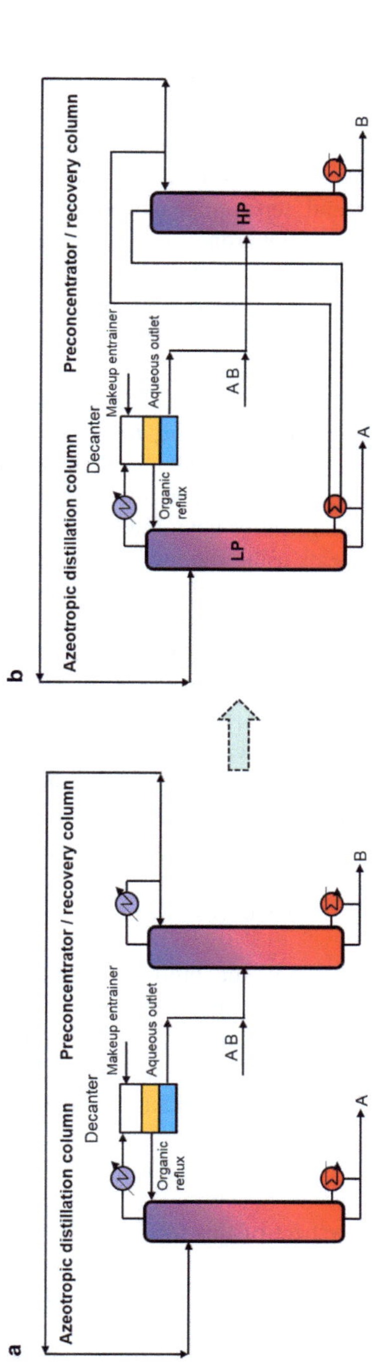

Fig. 5.22 Schematic diagram of (a) design 3 of heterogeneous azeotropic distillation sequence and (b) multi-effect heterogeneous azeotropic distillation sequence

reacting away contaminants, avoiding or eliminating difficult separations) while others are realized by using separation to improve reactions (e.g., enhancing overall rates, overcoming reaction equilibrium limitations, improving selectivity, removing catalyst poisons)—with the maximum effect being achieved when both aspects are important (Kiss, 2013). On the other hand, it should be considered using a reactive distillation carefully in some cases, such as incompatibility of process conditions for reaction and separation, or relative volatility constraints for reactants and products in the reaction zone of the distillation column, or the long residence time is required or catalyst contamination is a problem. Furthermore, when solid is present in the feed or when the reaction occurs in the gas phase, or when the use of enzymes as catalysts is required, the use of reactive distillation can cause many practical problems.

As there are several excellent books providing the details of steady-state design of reactive distillation columns, such as *Distillation, Principles and Practice* (Stichlmair and Fair, 1998), *Conceptual Design of Distillation Systems* (Doherty and Malone, 2001), *Reactive Distillation-Status and Future Directions* (Sundmacher and Kienle, 2003), and a comprehensive treatment of both steady-state design and dynamic control of reactive distillation systems using rigorous design approaches by commercial simulators, such as *Reactive Distillation Design and Control* (Luyben and Yu, 2008), there is no need to go over these topics in detail. The authors will only describe briefly reactive distillation and then move to process modifications, process integration, and process intensification technologies used to retrofit of reactive distillation.

As there are many types of reactions, there are many types of reactive distillation columns. In this section, the ideal classical situation is described, which will serve to outline the basics of reactive distillation. Consider the system in which the chemical reaction involves two reactants (A and B) producing two products (C and D). The reaction takes place in the liquid phase and is reversible (Luyben and Yu, 2008).

$$A + B \Leftrightarrow C + D.$$

For reactive distillation to work, the products should be removed from the reactants by distillation. This suggests that the products should be lighter and/or heavier than the reactants. In terms of the relative volatilities of the four components, an ideal case is when one product is the lightest and the other product is the heaviest, with the reactants being the intermediate boiling components.

$$\alpha_C > \alpha_A > \alpha_B > \alpha_D.$$

Figure 5.23 presents a flowsheet of this ideal reactive distillation column. The lighter reactant A is fed into the lower section of the column, while the heavier is fed into the upper section of the column (Luyben and Yu, 2008). The middle section is the reactive section to produce light product C and heavy product D, which are collected in the distillate and bottom streams, respectively. Note that according to the distribution of reactants and products in relative volatility ranking, the reaction zone can be located not just in the middle section, but also in the

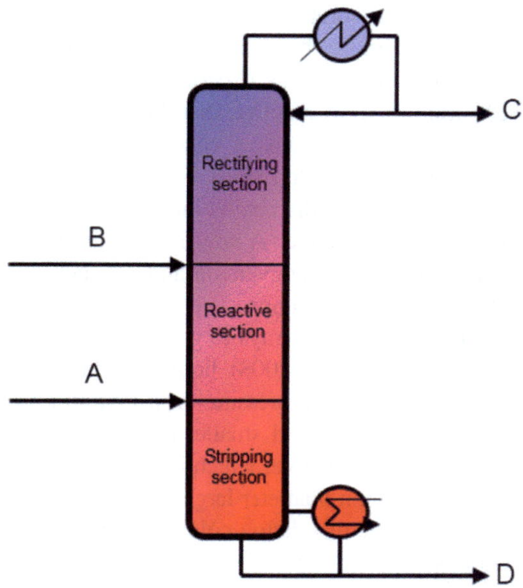

Fig. 5.23 Ideal reactive distillation column

top and bottom sections. This is because the reaction section should be located where the concentration of at least one reactant (preferably the limiting reactant) is the maximum (Subawalla and Fair, 1999).

Famous Eastman's first methyl acetate reactive distillation tower started up in 1980 (Agreda et al., 1990). It is more than 80m tall with a diameter of about 4m. Reactive distillations have been implemented on a commercial scale in the petrochemical industry more than 150 times (Harmsen, 2007). Normally, reactive distillations is applied to etherification, esterification, hydrolysis, and alkylation on an industrial scale with the largest number of installations is methyl tertiary butyl ether (MTBE), which is used in gasoline blending. Other esters are also produced now by reactive distillation, such as ethyl tert-butyl ether (ETBE), tert-amyl methyl ether (TAME) or fatty acid methyl esters (FAME). In recent years, considering the green opportunities, reactive distillation have gained considerable academic and industrial interest (Lutze et al., 2010; Kiss, 2013; Keller, 2014).

As compared to the design of a conventional distillation column, the design of reactive distillation is more complex with more specifications to accomplish both product compostions and reaction conversion. Those specifications include pressure-temperature, number of trays in stripping, rectifying and reactive sections, location of reactant feed streams, reactant ratio, catalysts selection, reaction volume provided by column internals and column diameter, reflux ratio, and reboiler duty. Many methods have been developed for the synthesis and design of reactive distillation columns and generally be classified into three broad categories: graphical, heuristic, and optimization methods (Almeida-Rivera et al., 2004).

Note that compared to the design of a conventional distillation column, it is much more sensitive to pressure due to the need for a match between the temperature favorable for the reaction and the temperature favorable for separation. Furthermore, holdup has no effect on the steady-state design of a conventional column, whereas it certainly affects dynamics design. However, holdup is very important in reactive distillation because reaction rates directly depend on holdup (or the amount of catalyst) on each tray (Luyben and Yu, 2008). This means that the holdup must be known before the column can be designed and before the column diameteris known. As a result, the design procedure for reactive distillation is iterative. A tray holdup is assumed and the column is designed to achieve the desired conversion and product purities. The diameter of the column is calculated from maximum vapor-loading correlations. Then the required height of liquid on the reactive trays to give the assumed tray holdup is calculated. According to Luyben and Yu (2008) liquid heights greater than 10–15 cm are undesirable because of hydraulic pressure-drop limitations. Thus, if the calculated liquid height is too large, a new and smaller tray holdup is assumed and the design calculations repeated. An alternative, which may be more expensive in terms of capital cost, is to make the column diameter larger than that required by vapor loading.

The reactive distillation column described above was operated in neat mode. The two reactants are fed in exactly the correct amounts to satisfy the stoichiometry of the reaction (Luyben and Yu, 2008). The control system must be able to detect any imbalance, which will inevitably result in a gradual buildup of one of the reactants and a loss of conversion and product purities. An alternative to operating neat is to operate the reactive column with an excess of one of the reactants. This eliminates the need to perfectly balance the reactant feeds in the reactive column itself, which makes the control of the reactive column easier. However, this mode of operation may have the disadvantage of requiring the recovery and recycle of the reactant that is in excess. The flowsheet typically consists of a two-column system: a reactive column and a recovery column. In addition, using more reactive trays can improve the controllability of a reactive distillation (Luyben and Yu, 2008).

Reactive distillation processes can be divided into homogeneously catalyzed reactive distillation process and heterogeneous processes catalyzed reactive distillation. Heterogeneous reactive distillation is often termed catalytic distillation to differentiate between the two processes (heterogeneous and homogeneous) (Subawalla and Fair, 1999). In general, homogeneously catalyzed reactive distillation processes are more flexible because the catalyst concentration and therefore the reaction velocity can be varied within a wide range (Keller, 2014). Furthermore, fresh catalyst can be continuously added to the reactive distillation column. However, homogeneous catalysis requires an additional separation step to recover the catalyst. This step can be avoided in heterogeneously catalyzed reactive distillation processes by special constructions that immobilize the catalyst within the column (Keller, 2014). Because of the catalyst immobilization, the position of the reactive zone can be precisely set within the column. However, exchanging the used catalyst is very work-intensive. Heterogeneously catalyzed reactive distillation is thus often avoided as a process option if a short catalyst lifetime would require too many shutdowns.

5.7.2 *Reactive Distillation Retrofit*

5.7.2.1 Column Internal Improvement

To improve the energy efficiency and/or reactant conversion of reactive distillation, in addition to the operating conditions, such as the column pressure, catalyst mass, reactant ratio, reflux ratio, there are several structural factors that affect reactive distillation such as the reactive zone location, reactant feed location, column diameter, number of equilibrium stages, packed height, holdup, and prereactor. Here, this book focused on determining the structure factors, which can be modified to achieve higher efficiency in terms of conversion, product selectivity, and energy. The main ideal is increasing the liquid holdup or amount of catalyst. As holdup is increased, vapor boilup and reflux steadily decrease. Thus, larger tray holdups reduce energy consumption. Another idea is replacing the existing column internal with high performance trays/packings. These ideas can allow an increase in reflux rate, which can enhance the conversion in equilibrim-limited reactions by giving steeper concentration profiles in the reaction section.

For homogeneously catalyzed reactive distillation processes, conventional column internals can be applied (trays or random or structured packings) (Keller, 2014). Because the reaction takes place in the liquid phase, the proper choice of column internals depends on the liquid holdup provided by the internals. In addition, the reaction velocity of the chemical system must be considered. In the case of fast reactions, the reaction reaches chemical equilibrium within the residence time range typical for column internals (Keller, 2014). Therefore, the column internals should be selected with regard to the separation efficiency. Consequently, random or structured packings are often used as column internals for fast, homogeneously catalyzed reactions. For slow reaction rates, high residence times and thus large liquid holdups must be provided by the column internals, and consequently tray columns are used (Keller, 2014). The achievable residence time increases from sieve to valve to bubble cap trays. The tray columns used for reactive distillation processes are operated in the bubbly flow regime rather than in the spray regime used in conventional distillation processes to ensure a greater liquid content and higher residence time for the chemical reaction. When the equipment used for homogenous reactive distillation processes consists of sieve trays, the height of the weirs that provide the necessary liquid holdup and residence time required for a reaction should be considered as an important design variable. In particular, a larger liquid holdup (larger amount of catalysts) and residence time can be adjusted by increasing the weir height. Note that liquid heights greater than 10–15 cm are undesirable because of hydraulic pressure-drop limitations even very large increases in holdup may improve controllability (Luyben and Yu, 2008). In addition, when the reactive zone is located in the bottom, this can be accomplished by removing some trays in the column bottom to increase the liquid level, which affects the reaction dramatically. In this case, the reboiler return location is also moved up, which should be designed by experienced engineers. In contrast, a new reflux drum can be installed in parallel with existing one to increase the liquid holdup when the reactive zone is located toward the top of reactive distillation column. Note that increasing the liquid holdup can improve the product amount and increase the conversion of unwanted by-products, which cause some loss in the reactants. Therefore, economic analysis needs to consider not only the investment cost to install new equipment or modification fee, operating cost in the purification unit, which can be reduced due to the smaller excess reactants, lost reactants cost and increased product cost.

Heterogeneous reactive distillation processes use a solid catalyst, where most of the reaction takes place within the catalyst particle. The packing and tray designs used for heterogeneous reactive distillation are substantially different from the conventional packing and tray geometries. These novel designs have two important features: they maximize the reaction efficiency by providing intimate contact between the vapor, liquid, and solid catalyst and simultaneously minimize pressure drop through the catalyst bed. In general, two different types of catalytic internals are available: structures that themselves are catalytically active or structures in which the catalyst pellets are immobilized (Keller, 2014). Two commercially used packing types are catalytic bale (Smith, 1980) and structured packing (DeGarmo et al., 1992). Other catalytic packing types, such as structured carpet packing (Pyhälahti and Klemola, 1998) and ring-shaped ion-exchange resins (Sundmacher and Hoffmann, 1993), have also been used. Recent developments of catalytically active structures, such as monolithic structures (Moulijin et al., 2011), and structures in which catalyst pellets, such as sandwich-type reactive packings Katamax® packing provided by Koch-Glitsch (Gelbein and Buchholz, 1991), Katapak-S® and Katapak®-SP packings manufactured by Sulzer Chemtech Ltd (Stringaro, 1995; Moritz and Hasse, 1999; Behrens et al., 2006), and catalytic tray, such as D+R tray developed by BASF SE (Adrian et al., 2000; Keller, 2014), are attractive and lead to low energy consumption. Replacing the existing column internal with recent developed packings/trays is expected to improve the performance of existing reactive distillation column.

5.7.2.2 Adding a Prereactor or a Side Reactor

When increase in the capacity of the process, in which reactive distillation is arranged forward of the distillation sequence, a prereactor can be installed upstream of the reactive distillation (Fig. 5.24a) to support for reactive distillation as well as increase the conversion of reactants or reduce the recycling of the unreacted reactants and decrease the heat duty of downstream distillation columns. Furthermore, for many reactive systems that are not severely equilibrium-limited, the reaction rates are high for a large fraction of the conversion range, and decline abruptly only when the compositions and temperatures approach equilibrium (Subawalla and Fair, 1999). For such cases, this can be an attractive option to reduce the reboiler duty in reactive distillation and the purification sequence. Normally, the operating conditions in reactive distillation, such as the operating pressure and temperature, which affect both chemical kinetics and vapor-liquid equilibrium, are chosen to balance

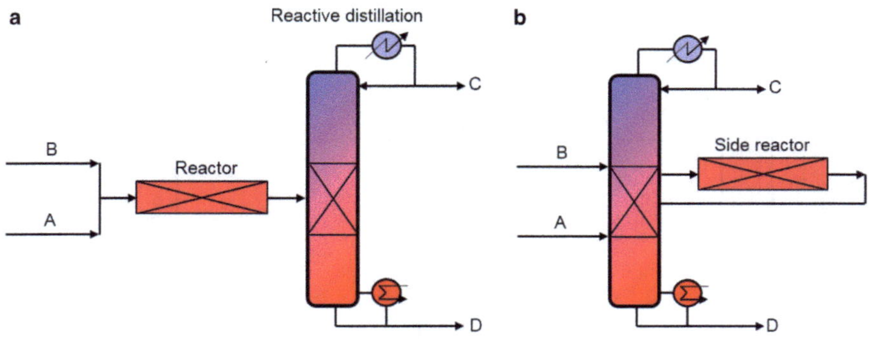

Fig. 5.24 Reactive distillation column with (a) a prereactor and (b) a side reactor

the reaction and separation. Therefore, another advantage of adding a prereactor is that the total reaction conversion can be enhanced by choosing the operating pressure and temperature for the reaction. For example, many methyl tert-butyl ether processes in industry use a prereactor in conjunction with a catalytic distillation, where the total conversion of the reactor-column pair approaches 100% (Hutchings et al., 1992; Subawalla and Fair, 1999). On the other hand, it might have the same problems related to by-product generation by increasing the liquid holdup.

Similar to adding a prereactor, side reactors can be installed in the reactive distillation (Fig. 5.24b) for high liquid or catalyst holdup to enhance the conversion and reduce the energy requirement in a separation train. This configuration can be also used when the temperature range suitable for reasonable chemical reaction kinetics does not match the temperature range suitable for vapor-liquid separation. The side liquid stream is pumped to a side reactor, in which the more of the reaction occurs, which is then fed back and might flash into the reactive distillation on the tray below. From a maintenance perspective, an external side reactor is preferred over an internal one. From a practical point of view, however, it is more difficult to install, operate, and control compared to adding a prereactor. Furthermore, the number of side reactors and their locations need to be optimized. Normally, the side liquid stream is withdrawn from the location in the column where the reactants are most abundant (Luyben and Yu, 2008).

5.7.2.3 Internal Heat Integration (associated with reaction zone)

The reactive zone depends on the reactant and product volatilities, which normally should be located where the concentration of at least one reactant (preferably the limiting reactant) is the maximum (Subawalla and Fair, 1999). In particular, this section is located toward the top of a column when the limiting reactant and product are the most volatile component and the least volatile component in the system, respectively. In contrast, the reactive section should be located in the bottom area when the product is the most volatile and in the middle section for a four-component system with two intermediate-boiling reactants. Therefore, to set an appropriate reactive

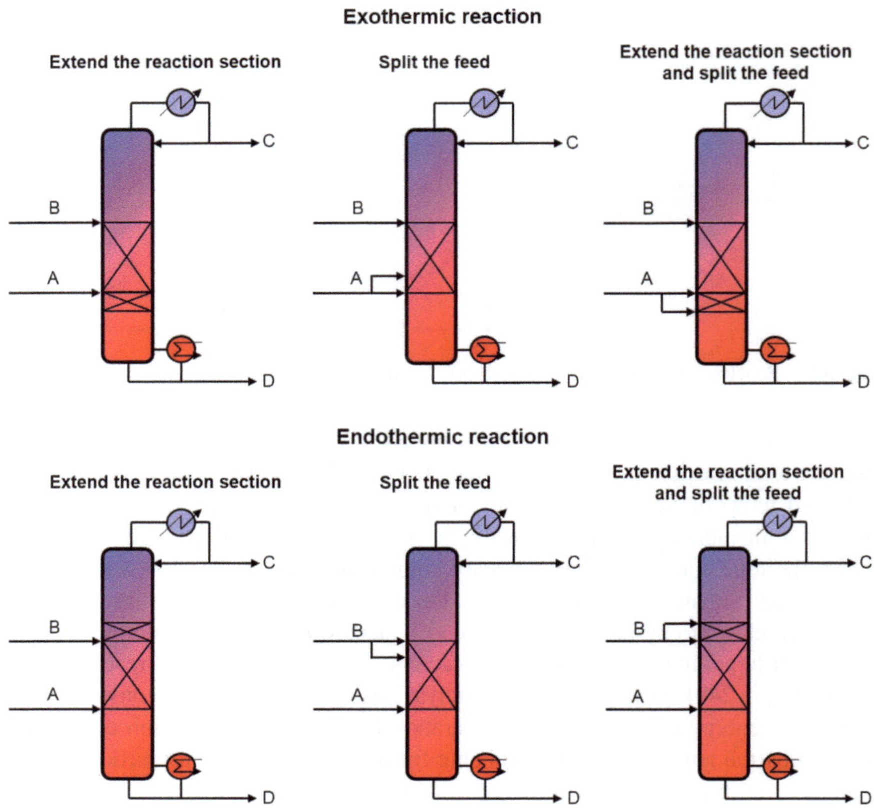

Fig. 5.25 Retrofit of a reactive distillation column using the internal heat integration concept

zone, many reactive systems are designed based mainly on this rule, while the internal heat integration, which utilizes the heat of reaction for separation, is not well recognized. The thermodynamic efficiency of a reactive distillation involving reactions with a highly thermal effect could sometimes be improved substantially by seeking further internal heat integration between the reaction operation and separation operation (Huang et al., 2005). For effectively conducting the process design for a reactive distillation column involving reactions with a highly thermal effect, it is often necessary to consider the internal heat integration by deliberately distributing the reactive section within the process.

For exothermic reactions, the reactive distillation column can be retrofitted by adding a solid catalyst in the packing or in the several trays below the reactive section to extend this, or splitting the feed into two feeds with one located at the same position and another fed to the tray above that feed, or using both of them (Fig. 5.25). Note that the new feed location can be located, the place where another component has high composition profile, which enhance the reaction. Similarly, for endothermic reactions, adding a solid catalyst in the packing or in the several trays above the reaction section, or splitting the feed into two feeds with one located at the same position and the other fed to the tray above that feed, or employing both of them, can accomplish the internal heat integration between the reaction operation and separation operation. By carrying out these methods, the reactive section can be extended to either the stripping section or rectifying section, which can improve not only the energy efficiency but also the reaction conversion, which can reduce further energy requirements in the downstream distillation columns that purify the product and recycle the unreacted reactants.

5.7.2.4 Feed Split

As discussed above, a feed split can enhance the reaction conversion because the partial feed of one component is spit and fed to the tray, where another component has a high value in the composition profile. Lee et al. (2010) reported that splitting the propionic acid in the production of butyl propionate can lead to significant savings in column diameter, which can result in a decrease in amount of catalyst required, and operating cost. In addition, a feed split can provide a chance for internal heat integration; however, making new feed nozzle might be difficult for a high pressure column due to strict government regulations.

5.7.2.5 External Heat Integration

To enhance the energy efficiency and/or capacity, all techniques applied for conventional distillation, such as feed preheating, self-heat recuperation technology, heat pump-assisted reactive distillation, or multi-effect reactive distillation, can be used in a reactive distillation column by utilizing the heat from the top vapor stream or bottom product. For example, a heat pump system can be used to intensify the reactive distillation (Fig. 5.26a). On the other hand, wide-boiling mixtures are more common than close-boiling mixtures. They can reduce the performance of heat pump due to the large temperature difference between the top and bottom section of the column. To avoid high-energy consumption, instead of transferring heat from the top vapor stream to the reboiler, that heat can be exchanged with a side reboiler (Fig. 5.26b). A side heat pump can be employed to reduce the vapor traffic in the top section and temperature difference (Fig. 5.26c and 5.26d).

5.7.2.6 Integration to TCRDS or RDWC

Reactive distillation has attracted considerable attention over the last two decades and has become an extensively researched process intensification method, because of its ability to increase the reaction conversion by removing products and simplifying the separation process by reacting the azeotropes away (Cheng at al., 2013). Normally, the technology of reactive distillation is applied for etherification, esterification, and alkylation on an industrial scale (Lei et al., 2005). In many cases, the reaction occurs in a single column, while some two-step reactions might take place in two columns.

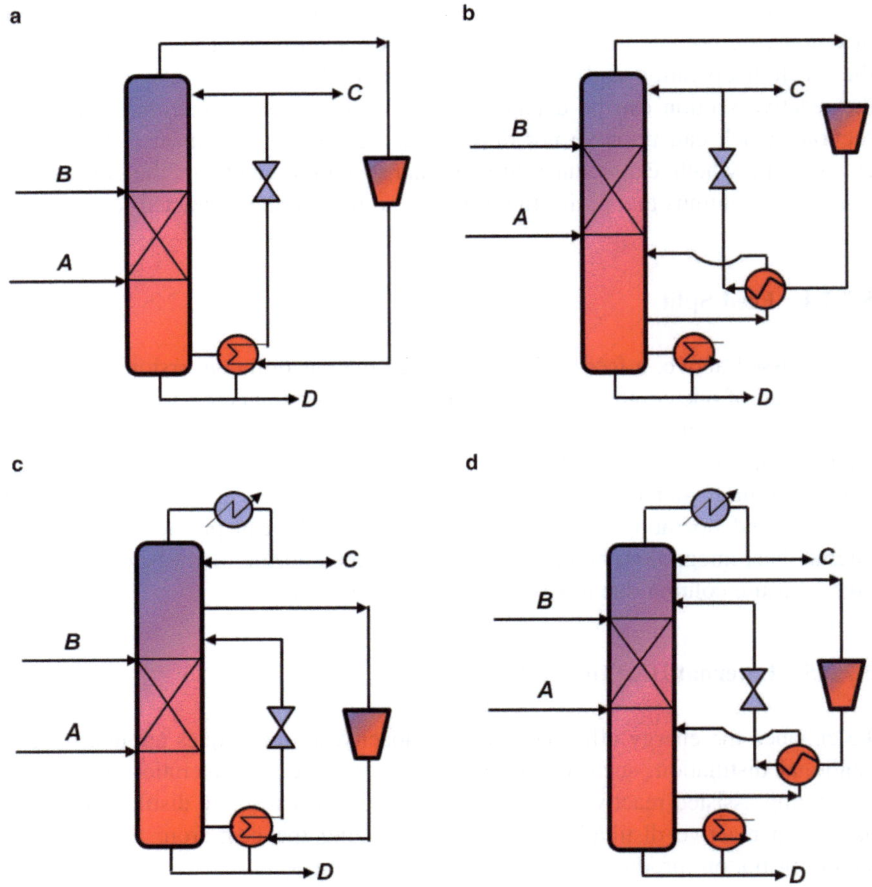

Fig. 5.26 Heat pump-assisted reactive distillation

Here, the integration of one reactive column with a conventional distillation column has attracted increasing attention, which can be extended to the integration of one separation column with one reactive distillation or two reactive distillation columns.

The thermal coupling concept, which improves the thermodynamic efficiency, can be employed to retrofit a reactive distillation. In particular, TCDS and DWC, which are the best examples of proven process intensification technology, have been proposed to form thermally coupled reactive distillation sequence (TCRDS) (Fig. 5.27) and reactive dividing wall column (RDWC) (Fig. 5.28), respectively, in an attempt to reduce the energy requirement and/or increase the capacity. Note that these sequences are used to integrate a reactive distillation and a separation column, while the sequences integrating a separation column with a reactive distillation or two reactive distillation columns might have different reactive section locations.

The pressure in the retrofitted TCRDS and RDWC can be adjusted to satisfy the operation or using water to condense the top vapor stream, which alters the

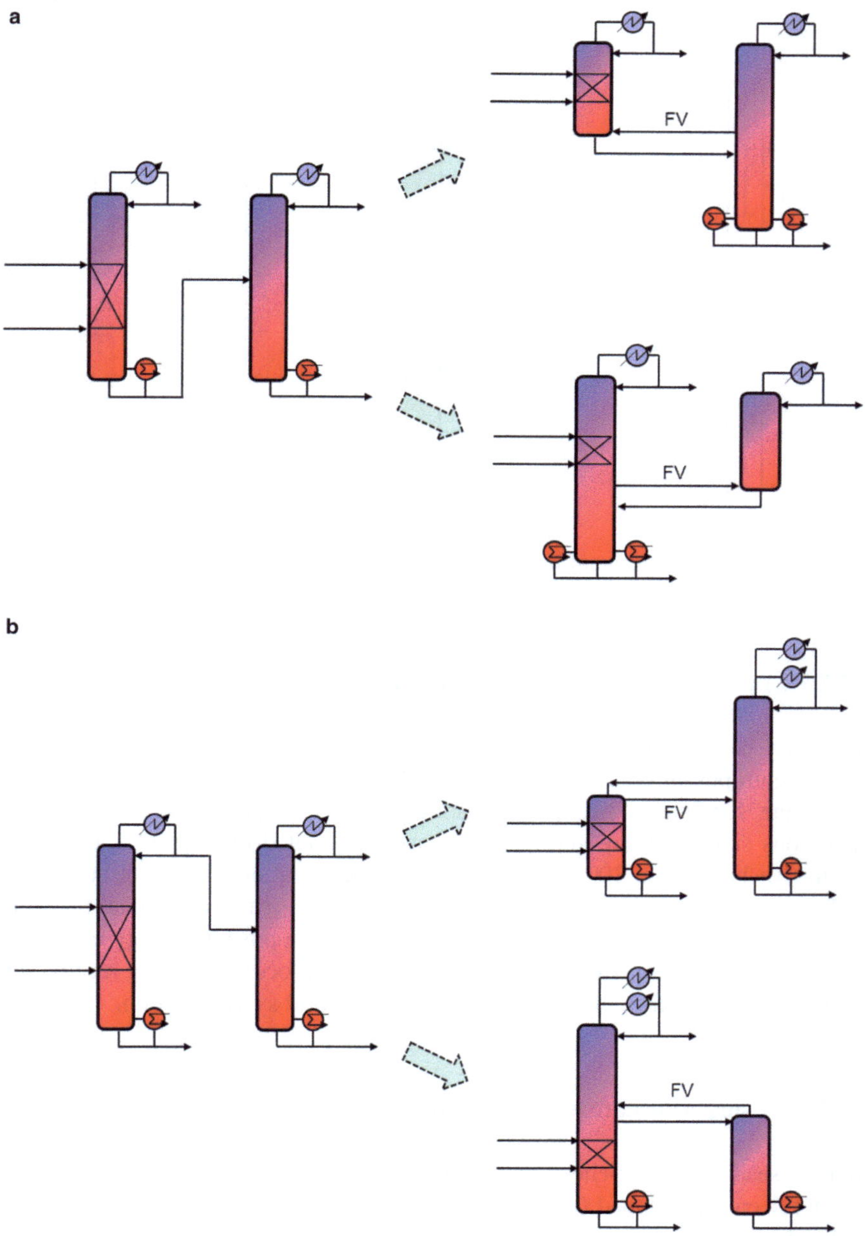

Fig. 5.27 Schematic diagram of the retrofitted TCRDC with (a) side rectifier and (b) side stripper

temperature profile in the reaction section affecting to the reaction rate and reaction equilibrium constant. In cases involving the integration of two reactive distillation columns, it has more constraints because two reactions might have different favorable operating pressures and temperatures.

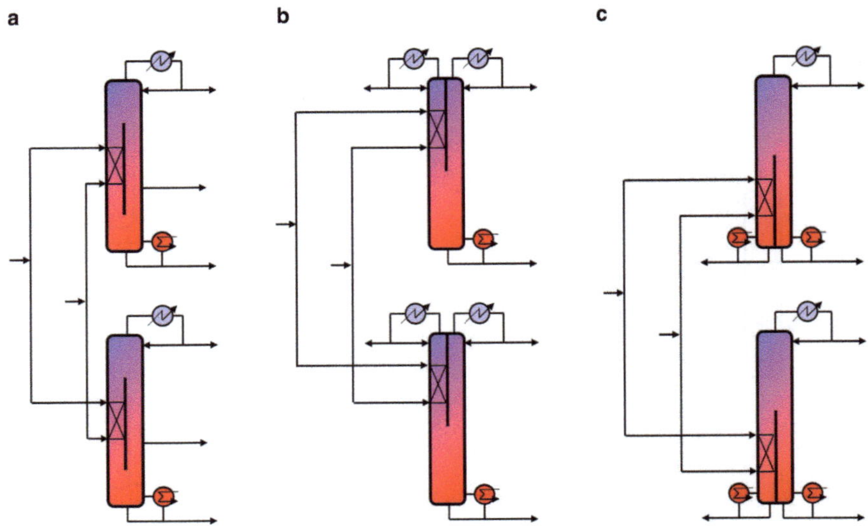

Fig. 5.28 Schematic diagram of the retrofitted (a) conventional reactive dividing wall column, (b) reactive top dividing wall column, and (c) reactive bottom dividing wall column

5.7.2.7 Integrating with Other Unit Operations

In several cases, non-ideal aqueous–organic mixtures are formed which tend to form azeotropes in reactive distillation columns, which can be overcome using membrane separations like pervaporation and vapour permeation or other unit operations since they are very selective and not limited by vapour–liquid equilibrium (Buchaly et al., 2007). Consequently, a new unit operation can be installed to form hybrid process, such as membrane-assisted reactive distillation, which contributes to sustainable process improvement due to arising synergy effects and allows reduction of operational costs of an existing process including reactive distillation column. This will be discussed more in hybrid system section.

5.7.2.8 Using Alternative Energy Sources

Many research groups believe that the potential of reactive distillation technology has not been fully tapped (Keller, 2014). Worth mentioning are the ideas to introduce energy into the reactive distillation through alternative energy forms such as microwaves and ultrasound (Altman et al., 2010; Lutze and Gorak, 2013). Because microwaves can improve both reaction and separation performance, their applications could accelerate the mass transfer and therefor may potentially enhance the performance of reactive distillation columns. However, recently, Werth et al. (2015) examined the impact of microwave irradiation on the separation and reaction in a reactive distillation column, but observed no apparent enhancement either in the separation efficiency on the macroscopic scale or in the reaction rate. Further study will be needed to prove the potential and feasibility of the concept of microwave-assisted reactive distillation.

Because retrofit of reactive distillation columns is in early stage, much effort needs to investgate and evaluate above process modification, integration and intensifiation technologies as well as to prove some concepts, such as using microwave in enhancing the performance of reactive distillation.

5.7.3 Case Study

5.7.3.1 Existing Plant Subject to Retrofit Study

In this example, one simple modification is employed to show the benefits when carrying out the retrofit of a reactive distillation column. In particular, the manufacture of methyl acetate, which is an important oil resin used in the manufacture of artificial leather, is considered to save operating costs. Figure. 5.29 presents the process flow diagram of the existing reactive distillation process (AEA Technology, 2000), which is used to produce and purify methyl acetate from methanol and acetic acid. The feed is fed onto the first reactive tray immediately above the stripping section. Note that this example has simple kinetic information, which is used mainly to illustrate the relative improvement of reactive distillation when retrofitted using a heat pump. Table 5.1 shows the kinetic parameters for the base case. The Wilson activity model was used to describe the thermodynamic properties. The base case simulation model showed that the energy consumption of the reactive distillation was 870 KW.

5.7.3.2 Heat Pump Assisted Reactive Distillation

Figure 5.30 shows the CGCC for the reactive distillation. An inspection of the CGCC confirmed that the column energy efficiency can be improved using a heat pump operating between the top and bottom with relatively small differences in

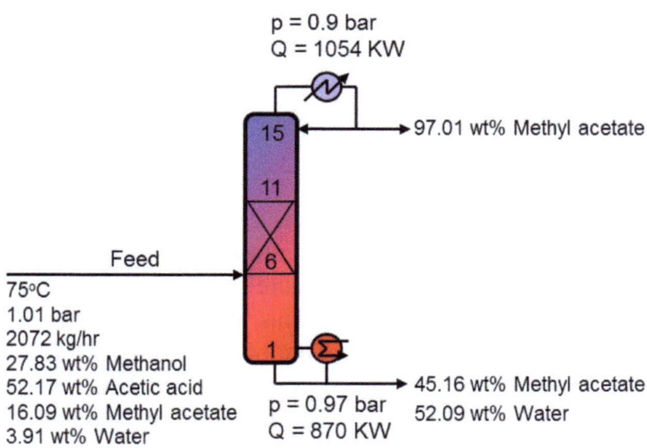

Fig. 5.29 Simplified flow sheet illustrating the reactive distillation

Table 5.1 Kinetic parameters for base case

Base component	methanol
Reaction phase	liquid
Forward reaction	
Arrhenius parameter A	1.0e5
Activation energy E (kJ/kgmole)	2.3e4

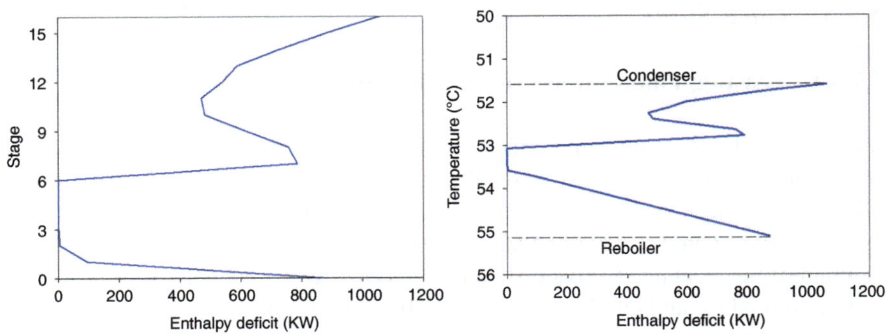

Fig. 5.30 CGCC for a reactive distillation column

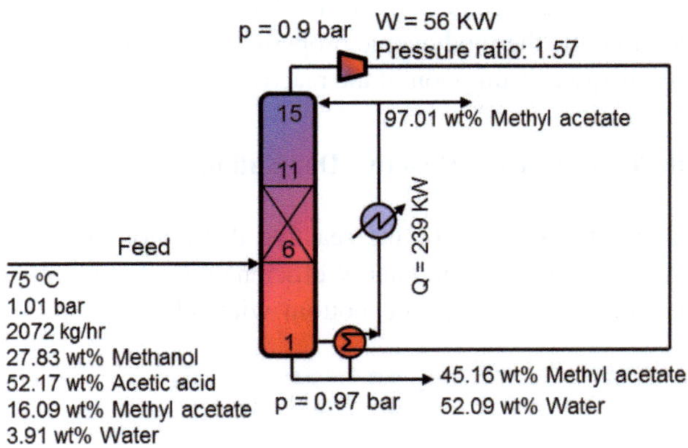

Fig. 5.31 Simplified flow sheet illustrating the retrofitted distillation with mechanical vapor recompression heat pump

temperature. As the heat load is high, there is a strong likelihood of improving the energetic efficiency of the system. Vapor from the top of the reactive distillation column was compressed to a certain pressure (increasing its vapor temperature) and condensed in the column's reboiler. This allows the heat of the condensing vapor to assist in vaporization at the bottom of the column via a heat exchanger. The top outlet stream must be cooled further before being divided into two streams: one that is recycled back to the column as reflux, and the other that is the final top product (Fig. 5.31). The pressure of the outlet compressor was adjusted to obtain the minimum approach in the heat exchanger of 10 °C. Compared to the conventional column sequence, the use of a top vapor recompression heat pump can reduce the condenser duty, reboiler duty and operating costs by 77.3%, 100.0% and 91.4%, respectively, compared to the conventional sequence.

5.8 Hybrid System

Interest in hybrid distillation processes is growing due the increasing demand for energy efficient processes in the chemical and biochemical industries (Roth et al., 2013; Nhien et al., 2017a). Especially when separating non-ideal multi-component mixtures into pure components, the smart combination of existing distillation processes with other separations can overcome existing limitations and bring about substantial synergies (Gorak and Stankiewicz, 2011; Moulijn et al., 2008; Long et al., 2016). Any single type of separation process is subject to limitations that can be overcome using intensified hybrid designs (Beneke et al., 2013). The difference between a hybrid process and a simple sequence of unit operations is illustrated in Fig. 5.32 (Ritter et al., 2012). In contrast to a simple sequential connection of different unit operations, hybrid separation processes are characterized by a mutual interdependence of different unit operations and overcome the limitations of the single unit operations by exploiting synergetic effects (Skiborowski and Gorak, 2016).

To select a suitable hybrid system for a given separation task, there are many ways. The trial-error experimental and/or simulated approach would be reliable but could time consuming and expensive (Babi and Gani, 2014). A rule-based selection procedure (Babi and Gani, 2014), which is required to upgrade for considering many other hybrid systems, and a model-based approach (Babi and Gani, 2014; Skiborowski and Gorak, 2016) can be applied to efficiently screen for suitable process configurations. Even model-based approach is time consuming and expensive, it is more accurate; and work is done at the end to obtain the final design. Although there are a lot of processes classified as hybrid separation processes, three most prominent processes are introduced and discussed in following sections.

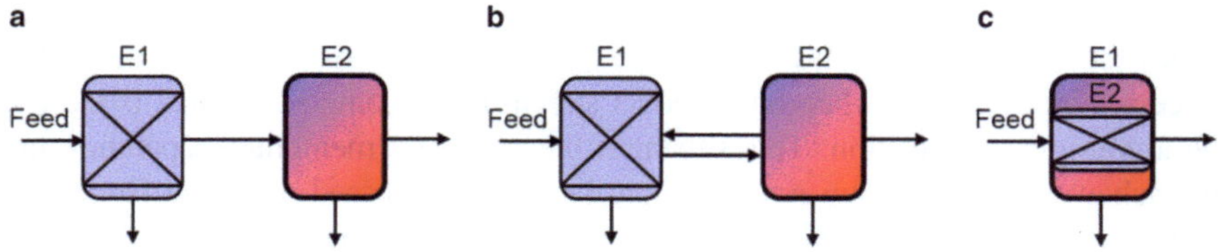

Fig. 5.32 Depictions of (a) a sequence of separation units, where the performance of E2 does not affect the performance of E1 (not a hybrid process), (b) a sequence of separation units, where the performance of E2 does affect the performance of E1 (a hybrid process), and (c) separation units, E1 and E2, residing in the same vessel (a hybrid process)

5.8.1 Hybrid Distillation-Membrane System

5.8.1.1 Introduction

In recent years, among the various separation processes available, membranes have attracted substantial interest from industrial and academic perspectives (Drioli et al., 2012; Sirkar et al., 2015; Long et al., 2016). The intrinsic properties of membrane operations make them ideal for industrial production: they have high selectivity and permeability for the transport of specific components; they are easy to scale up, modular, and generally athermal; they do not involve phase changes or chemical additives; and they usually have low energy consumption, the potential for more rational utilization of raw materials, and the potential for the recovery and reuse of byproducts (Drioli et al., 2012; Macedonio and Drioli, 2017). For these reasons, membrane engineering meets the requirements of green process engineering toward the realization of sustainable industrial development. Moreover, the integration of different membrane operations in the same industrial cycle can lead to further important benefits in terms of product quality, plant compactness, environmental impact, and energy use. Besides, various new membrane processes or membrane-based processes, such as membrane distillation, membrane absorption, membrane extraction, membrane reactor, with many advantages are either commercialized or being commercialized (Sirkar et al., 2015). Thus, membrane engineering has the potential to contribute towards process intensification by replacing the conventional energy-intensive separation techniques, such as distillation and evaporation, to achieve the selective and efficient transport of specific components, to improve the performance of the reactive processes, and to provide reliable choices for environmentally friendly industrial growth (Drioli and Curcio, 2007).

The hybrid configuration of distillation and membrane separation, which is normally used to separate azeotropic mixtures as well as close-boiling mixtures, has been one of the most promising process concepts (Gorak and Stankiewicz, 2011; Lutze and Gorak, 2013). It is because using a single conventional distillation column is impossible or expensive. It is also because if membrane processes are used alone, high membrane area is required. Combination of distillation and membrane can exploit synergetic effects between the single unit operations. These hybrid configurations allow more flexible process design and optimization. To facilitate the construction of a hybrid design, it is important to sequence the separation units such that each unit operates in the region of the composition space where its separation ability is best suited and most effective (Beneke et al., 2013). In particular, distillation processes are restricted when one product stream has a composition that is close to the azeotropic composition. This stream is then fed to a membrane separation unit that achieves the separation not based on the VLE, but based on the differences in diffusion rates through a solid membrane (Luyben and Chien, 2010). Because driving force for separation in a membrane process is different from that in distillation, it does not exhibit the same azeotropic behavior. Therefore, the limitations of distillation processes can be overcome using a membrane unit (Beneke et al., 2013).

Among the many types of membrane, pervaporation (PV) and vapor permeation (VP) are quite suitable for separating close boiling mixtures and in breaking azeotropic mixtures where distillation is energy intensive or requires the use of an auxiliary agent (Fontalvo et al., 2005). The main difference between PV and VP is that the feed in VP is supplied as a vapor, whereas in PV, the feed components change their aggregate state from liquid to vapor while permeating through the membrane (Kreis and Górak, 2006). In PV, the energy to vaporize the permeate is provided by the liquid feed stream. Therefore, the liquid stream exits the membrane module at a decreased temperature. A range of configurations can be found, where membrane module can be located in the column feed (Fig. 5.33a) (Verhoef et al., 2008), or within the column for breaking the distillation boundaries (Fig. 5.33b) (Fontalvo and Keurentjes, 2015), or in the side stream of the column for the separation of close boiling mixtures (Fig. 5.33c) (Kreis and Górak, 2006), or on the top (Fig. 5.33d) and bottom (Fig. 5.33e) of the column to perform the final product stream purification (Stephan et al., 1995). The technical viability of the hybrid distillation–membrane depends mainly on the permeability of the target compound through the membrane and the selectivity that the membrane can achieve (Luis et al., 2014). A high permeability means high production and a high selectivity refers to high purity, thus, an ideal membrane would have a high permeability and a high selectivity. However, the trade-off between permeability and selectivity makes it difficult to find the appropriate membrane. Note that in most cases, membranes are selected through expert advice and/or database search. In addition, the CAPEC database contains a small section for data on membranes, where reported data on membrane assisted operations have been stored (Babi and Gani, 2014).

These hybrid configurations, however, are complex in process design and optimization, and difficult to operate caused by the complex interactions between the combined unit operations (Holtbruegge et al., 2013a). The proper design of hybrid distillation-membrane system require knowledge of the phase and transport level, by introducing another phase, and the fundamental and molecular level, by means of selecting material (Skiborowski and Gorak, 2016). Furthermore, the short lifetime of membranes and the lack of process knowledge are also disadvantages (Kreis and Górak, 2006). Besides, membrane fouling, which is the accumulation of unwanted materials on the surface or inside the pores of membrane, is still one of the major issues that hounds the long-term stability performance (Tijing et al., 2015).

5.8.1.2 Retrofit Using Hybrid Distillation-Membrane System

The pressure swing or extractive distillation sequences are normally employed to separate azeotropic mixtures. Not all chemical plants use two columns in these sequences, some use only one column to obtain one main product, while another stream is recycled to the tank for treatment or sale to other plants or other reasons. On the other hand, it can waste raw material. Therefore, they are required to carry

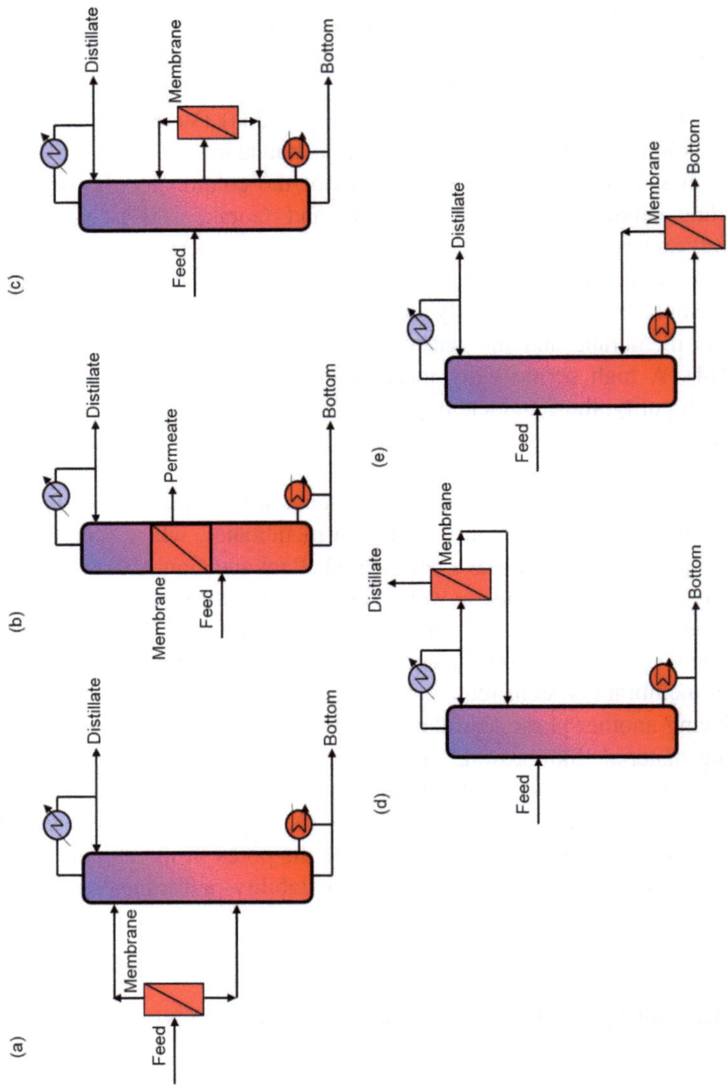

Fig. 5.33 Various configurations for a membrane/distillation hybrid process

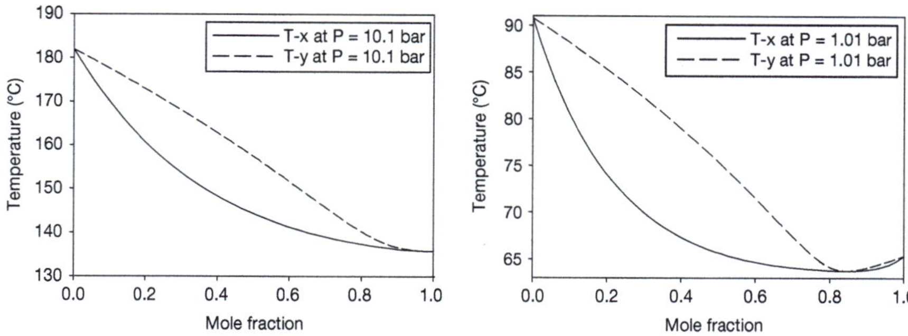

Fig. 5.34 Pressure sensitive, minimum boiling azeotrope

out a retrofit to increase the recovery of that material, particularly when it is expensive and requires a transportation fee.

For example, one high-pressure column is employed to separate dimethyl carbonate (DMC) and methanol mixture, which has a homogeneous azeotrope at 96% methanol (Fig. 5.34). On the other hand, column owners want to increase the DMC recovery from the top stream. As shown in Fig. 5.34, the composition of the azeotrope in this system changes according to the pressure, which shows that a low-pressure column can be installed to form a pressure swing sequence (Wang et al., 2010) (Fig. 5.35). Integration between these two columns is favorable because there is a temperature difference between the top of the high-pressure column and the bottom of the low pressure column. Furthermore, adding an extractive solvent, such as aniline or phenol, can enhance the relative volatility and installing a new column can form an extractive distillation sequence or TCEDS-SR (Hsu et al., 2010, Long and Lee, 2013b) (Fig. 5.35).

Using these sequences is very challenging and cost-intensive. Recently, Holtbruegge et al. (2013b) reported that it is possible to separate an azeotropic mixture of DMC and methanol using a hydrophilic VP membrane. Holtbruegge et al. (2013a) used this type of membrane to assist in reactive distillation, where the top vapor stream comprising of DMC and methanol in the partial condenser is fed to a hydrophilic VP membrane. Owing to the affinity of the membrane for hydrophilic methanol, the permeate consisted mainly of the preferentially permeating component methanol, while the retentate contained the retained DMC. This membrane can be used in the retrofit of that high-pressure column to improve the DMC recovery (Fig. 5.35).

Pervaporation is one of the most promising technologies for debottlenecking purposes since it is not affected by the presence of an azeotrope (Genduso et al., 2016) and because pervaporation is a low energy consumption technology (Lipnizki et al., 1999). Typically, the upstream side of the pervaporation is at ambient pressure and the downstream side is under vacuum to allow the evaporation of the selective component after permeation through the membrane. In pervaporation the major challenges can be expensive because of relatively low

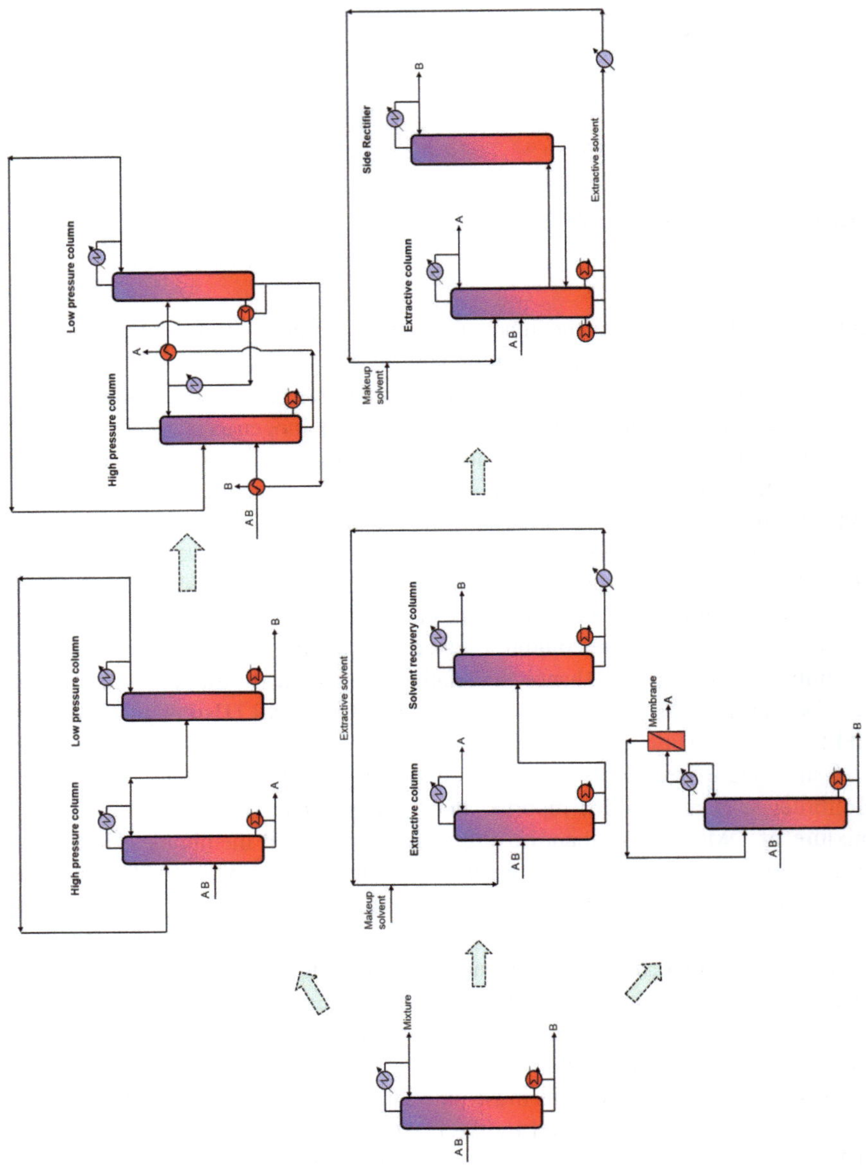

Fig. 5.35 Schematic diagram of the retrofitted sequences to improve the DMC recovery

permeate fluxes and the condensation of organics of high volatility, since very low refrigerant temperatures may be required. Debottlenecking of a isopropyl alcohol (IPA) dewatering unit by a membrane module was studied by Van Hoof et al. (2004). From the results it can be concluded that the hybrid process consisting of distillation followed by pervaporation with ceramic membrane is economically the most interesting process that leads to a decrease of the total costs of approximately 49%. Recently, a pervaporation module was employed in order to retrofit the extractive distillation unit used to recover methyl acetate from a methanol/methyl acetate industrial wastestream (Genduso et al., 2016). The retrofitting permits to avoid the use of a solvent during azeotropic separation. Moreover, during comparison of the retrofitted hybrid pervaporation-distillation unit with the source industrial extractive distillation unit, an overall energy saving up to 38% was demonstrated. In addition, major hydraulic modifications could be avoided employing structured packing and membranes of higher separation factors.

5.8.1.3 Retrofit Using Hybrid Membrane-TCDS or Hybrid Membrane-DWC or Hybrid Membrane-RDWC Systems

The thermal coupling concept, which improves the thermodynamic efficiency, can be employed to combine with a membrane to retrofit two-column sequences. In particular, TCDS and DWC, which are the best examples of proven process intensification technology, are proposed to form hybrid membrane-TCDS systems (Fig. 5.36) and hybrid membrane-DWC systems (Fig. 5.37), respectively, in an attempt to overcome the azeotrope, reduce the energy requirements and/or increase the capacity and, sustainability.

In reactive distillation, the product and unreacted reactant sometimes form azeotropic mixtures, which can be broken by a membrane. The theoretical and experimental investigation showed that a hybrid process of reactive distillation and membrane is feasible (Holtbruegge et al., 2013a; Buchaly et al., 2007). The coupling of a membrane separation unit with a reactive distillation column contributes to sustainable process improvement due to arising synergy effects and allows reduction of investment and operational costs (Buchaly et al., 2007). Furthermore, the novel configuration of membrane-assisted reactive dividing wall column or hybrid membrane-RDWC system (Fig. 5.38), which combines the reactive distillation used to overcome the chemical equilibrium, and DWC employed to save energy requirement, with the membrane utilized to overcome the azeotrope, is proposed in the retrofit (Long et al., 2016). Note that the unreacted reactant separated by a membrane can be withdrawn or recycled to the RDWC. This intensified configuration is also can be considered in the new plant design. More detailed research will be needed to understand better this process as well as to quantify the synergistic effects of the three unit operations. Insight analysis should be performed, which can be used to define the operating range of this novel process.

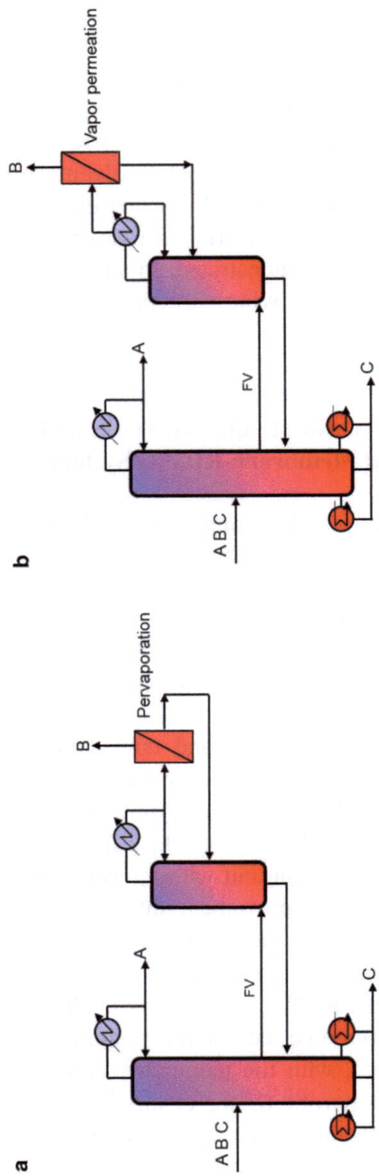

Fig. 5.36 Schematic diagram of the retrofitted hybrid membrane-TCDS systems

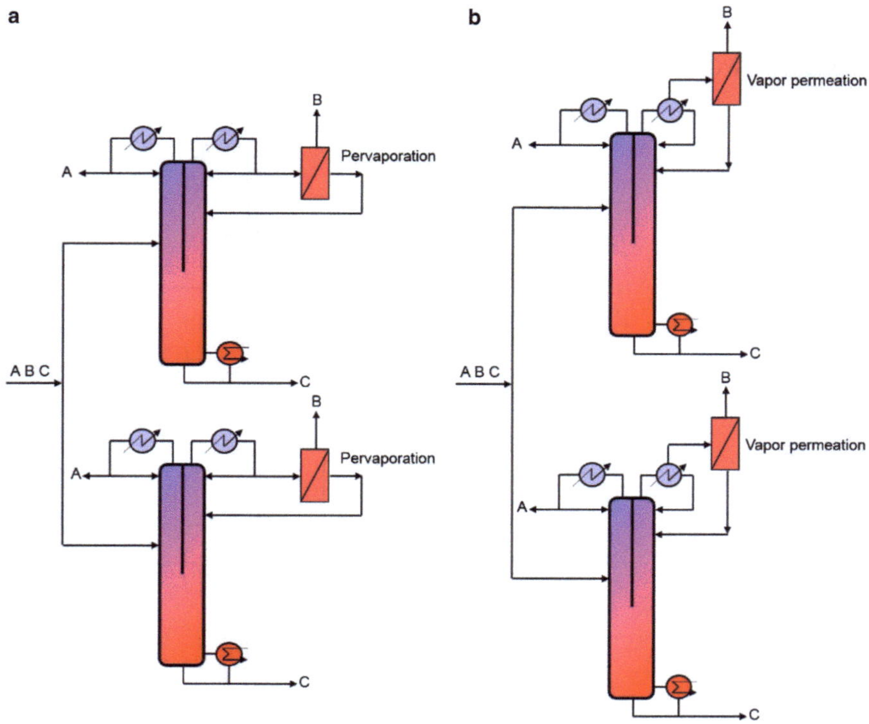

Fig. 5.37 Schematic diagram of the retrofitted hybrid membrane-DWC systems

5.8.2 *Hybrid Distillation-Extraction System*

5.8.2.1 Introduction

Normally, separation of mixtures containing a large quantity of water by distillation is very energy intensive because of the presence of a heterogeneous azeotrope between target components and water, which can occur in biofuel and biochemical production from biomass (Nhien et al., 2016a,b). In those cases, a hybrid purification process that combines extraction and distillation (shown in Fig. 5.39a) to purify target components can be considered and investigated in detail to improve the process efficiency of existing conventional distillation sequence. In many cases, the extraction cannot achieve high purity product, which is needed to further remove remained water and obtain other byproducts. In those cases, more than one distillation column are employed to products purification and solvent regeneration (as shown in Fig. 5.39b). The operating temperature during extraction is often far lower than that for distillation. Thus, the use of extraction can prevent the loss of products due to polymerization (Jong and Marcotullio, 2010). Solvent selection for extraction and the entire hybrid process is critical for process design and minimizing the production cost. The selection of a suitable solvent is challenging because of the different chemical natures of the species. The

Fig. 5.38 Schematic diagram of the retrofitted hybrid membrane-RDWC systems

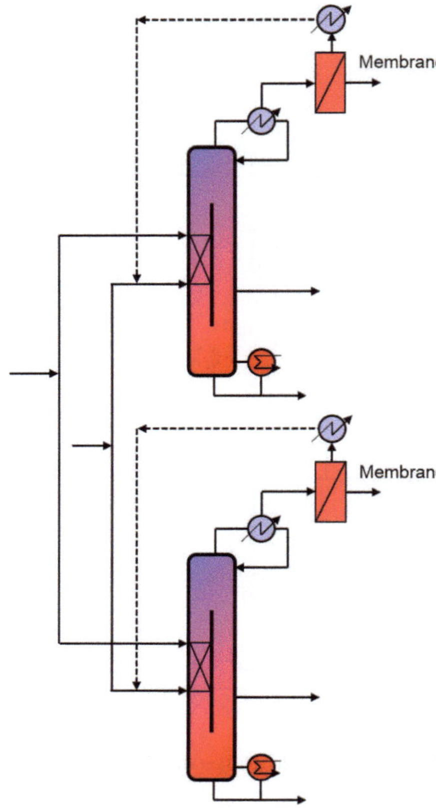

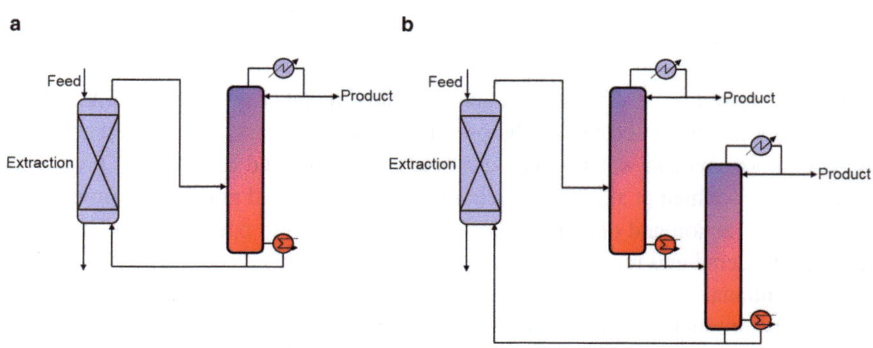

Fig. 5.39 Schematic diagram of the hybrid distillation-extraction systems

solvent must satisfy the following: equilibrium distribution, selectivity, density, boiling point temperature, viscosity, toxicity, cost, and ease of regeneration.

Several methodologies for solvent selection have been proposed (Gebreslassie and Diwekar, 2015). The first approach uses traditional laboratory synthesis and test methodology to find promising solvents. Even though this method may be limited by cost, safety, and time, the results achieved from this are accurate and reliable. The other approaches, based on screening the property database and computer-aided molecular design (CAMD), can be used for the efficient and reliable design of candidate solvents from their fundamental building blocks. The heuristic numeration strategies, mixed integer nonlinear programming model, were applied to solve CAMD techniques (Li et al., 2002; Karunanithi et al., 2005). While the CAMD methodology facilitates an efficient screening of a vast number of potential solvents, the accuracy of the property predictions limits the power of CAMD and necessitates experimental validation (Skiborowski and Gorak, 2016). Furthermore, even if a solvent has outstanding solvent power and selectivity it might be difficult to recover, resulting in an inefficient process.

Thus, an important criterion of solvent selection is the cost of the entire hybrid extraction/distillation process, which includes extraction and the distillation sequence for purification as well as solvent regeneration (Nhien et al., 2016c). This makes the problem more complicated and requires a new, simple, and efficient method to select a suitable solvent not only for zeotrope mixtures but also for azeotrope mixtures. Figure 5.40 shows a proposed systematic framework of solvent selection. This flowchart essentially has nine steps. In the first step, the

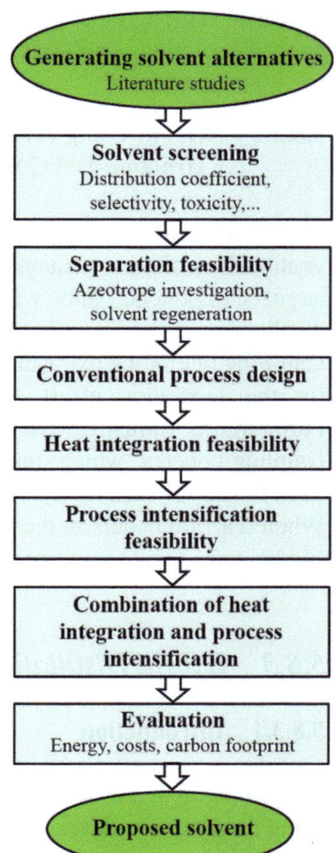

Fig. 5.40 Systematic procedure for solvent selection

potential solvents are identified either by previous studies or heuristic rules. Next, the solvents selected are screened carefully on the basis of their equilibrium distribution, selectivity, toxicity, and others. The distribution coefficient of the extracted component (K_D) is the ratio of the weight percentage (wt.%) of the extracted component in the solvent phase to the weight percentage of the extracted component in the aqueous phase (Demesa et al., 2015).

$$K_D = [\text{extracted component}]_{\text{solvent}} / [\text{extracted component}]_{\text{aqueous}} \quad (5.1)$$

The selectivity is used when there are more than one solutes (e.g., solutes A and B). The selectivity between the two solutes And B is defined as the ratio of the distribution coefficient of A to the distribution coefficient of B. In the third step, the separation feasibility for each solvent is examined quickly by investigating azeotropes and the ease of solvent regeneration. The processes of selecting candidates are detailed in the next step. Then feasibilities of heat integration, process intensification, and combination of them for improving the conventional process are respectively analyzed and evaluated. The total energy consumption, annual cost, and amount of carbon dioxide emission of each process are then calculated. Finally, the potential solvent are proposed on the basis of the overall performance. Through these steps, the potential solvents, the optimal column designs, and the operating conditions can be investigated and evaluated.

5.8.2.2 Retrofit Using Hybrid Extraction-TCDS or Hybrid Extraction-DWC Systems

The hybrid extraction-distillation system can be used to retrofit the existing conventional distillation sequence in separation of heterogeneous azeotrope between target components and water. Because the energy demand of any hybrid distillation-extraction process is determined by solvent recovery, the minimum solvent flow rate in the extraction column has been assumed to be a suitable indicator for the separation effort of the hybrid process (Skiborowski et al., 2013). Furthermore, similar in case of hybrid membrane-distillation systems, the thermal coupling concept, which improves the thermodynamic efficiency, can also be utilized to retrofit existing hybrid extraction-distillation systems (shown in Fig. 5.41). When reaction occurs in the process, reactive extraction and/or reactive distillation can be considered to improve the existing process.

5.8.3 Hybrid Distillation-Adsorption System

5.8.3.1 Introduction

For removing small quantities of targeted components from a mixture, adsorption using molecular sieves is popular because of its selectivity and ease of

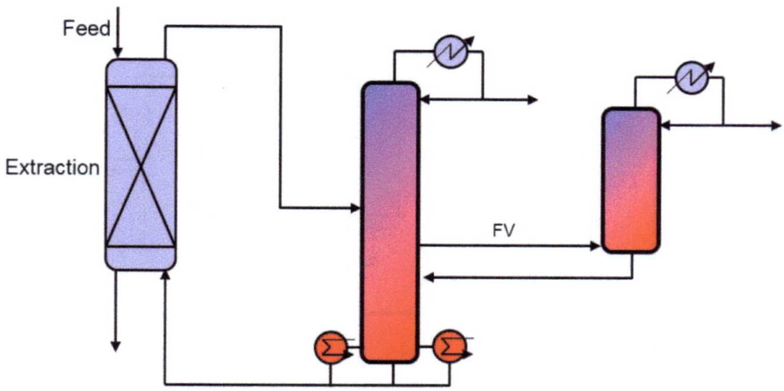

Fig. 5.41 Schematic diagram of the retrofitted hybrid extraction-TCDS system

maintenance (Loy et al., 2015). Adsorption is a separation process that works on the molecular level. One of its mechanisms involves the entrapment of target molecules in pores by the formation of temporary bonds so that they can be separated from other molecules. Making use of the difference in molecular size, pore sizes are made such that molecules larger than the target molecules cannot enter. For example, the 3A zeolite molecular sieve, which has a nominal size of 3 Å, is most commonly used for dehydration of ethanol since water molecules with an approximate molecular diameter of 2.8 Å, can easily penetrate the pores of the molecular sieve adsorbent, while ethanol, with an approximate molecular diameter of 4.4 Å is retained (Carmo and Gubulin, 1997; Venkatesan, 2013). Therefore, ethanol molecules cannot enter 3 Å pores and pass right over the molecular sieves.

Strength of adsorption of target molecules in molecular sieves depends strongly on temperature and pressure, thus giving rise to the two processes of temperature swing adsorption (TSA) and pressure swing adsorption (PSA) (Loy et al., 2015). The molecules are strongly adsorbed at high pressures and low temperatures, and the molecular sieves are regenerated at low pressures and high temperatures, respectively (Sowerby and Crittenden, 1988). The fast changes in the pressure enable short cycle times and thus much higher throughput is possible in the case of the PSA compared to the TSA systems. In addition, the TSA is energy intensive because of the need to supply heat for the regeneration. Thus, PSA is normally preferred to TSA.

One of the promising hybrid processes combines adsorption or PSA with distillation (as shown in Fig. 5.42). The design and modelling of such a process requires a good understanding of the characteristics of both distillation and adsorption. It is advantageous if the adsorption portion of the hybrid process is carried out at the same or similar operating pressure of distillation (Ghosh et al., 1993). A new approach for modelling hybrid PSA-distillation process was developed (Ritter et al., 2012). The approach involves two parts. Part I determines if energy savings

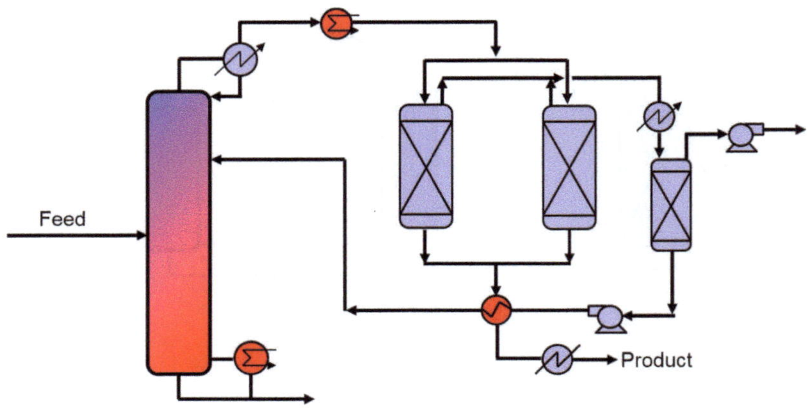

Fig. 5.42 Schematic diagram of the hybrid distillation-adsorption system

are possible. It can be done easily with sufficient knowledge of distillation process design, but with only minimal knowledge of PSA process design. Part I is carried out using a distillation process simulator to model a distillation column connected to a PSA unit that is treated as a "black box" with an assumed process performance. In this way, a hybrid PSA-distillation process can be analyzed simply by performing mass balances around these units and running simulator to determine if energy savings are possible compared to an existing process. Once an energy savings hybrid "black box" PSA-distillation process is found in part I, part II determines if an "actual" PSA process exists that mimics its performance. Part II is carried out using a rigorous PSA process simulator such as Adsim from AspenTech; thus, it requires significant knowledge of PSA process design. The outcome of part II is a hybrid PSA-distillation process that has the potential to be more energy efficient than the existing process. This new methodology should be very useful for quickly accessing the utility of hybrid PSA-distillation processes for a variety of other applications, with many possibilities for achieving significant energy savings and/or throughput debottlenecking.

5.8.3.2 Hybrid Distillation-Adsorption System Retrofit

The production of cellulosic ethanol (CE) from lignocellulose was developed at National Renewable Energy Laboratory in the US (Humbird et al., 2011). The lignocellulosic biomass type used in this process was corn stover, which is the most abundant agriculture residue in the US. Basically, the production of CE has three main steps: pretreatment of raw materials, bioethanol fermentation, and separation. Among them, the energy-intensive separation process comprises a major proportion of the overall production cost (Huang et al., 2008). Owing to the presence of an azeotrope between ethanol and water (95.63 wt.% ethanol), the downstream of

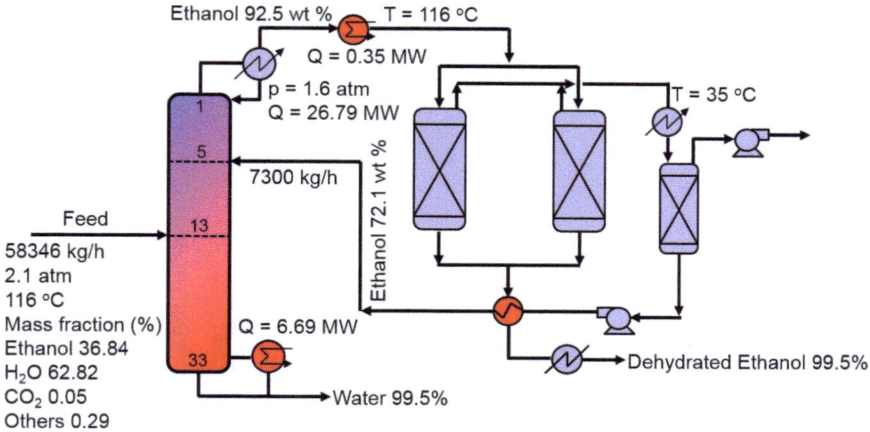

Fig. 5.43 Schematic diagram of the hybrid distillation-adsorption system for CE production

separation process usually includes distillation column producing water at bottom and the close azeotropic composition mixture between water and ethanol as the top vapor stream (Nhien et al., 2017b) and adsorption employed to break the azeotrope to achieve anhydrous ethanol (99.5 wt.%) (as shown in Fig. 5.43).

Vapor from the top of the distillation can first be divided into two streams. One stream is compressed to increase its vapor temperature and then condensed in the column's reboiler and feed heater of the adsorption though indirect contact with the liquid in the column's bottom and adsorption feed, respectively (Fig. 5.44). Compared with the conventional column sequence, using a top vapor recompression heat pump can reduce energy consumption by up to 21.5%, 100.0% and 100.0% in the condenser, reboiler, and feed heater duty, respectively, bringing about a 77.3% reduction of operating costs.

The hybrid distillation-adsorption system can also be applied to the recovery of organics from aqueous mixtures (as shown in Fig. 5.45) (Mersmann et al., 2011). Most of the water is recovered as bottoms in column C-1, while the overhead fraction with azeotropic concentration is fed, in vapor state, into the adsorber for removal of the residual water. Pure organics are withdrawn from the adsorber. The organics/water mixture generated in the regeneration mode is fed into column C-2 for further processing. In some cases, the latent heat of the vaporous organic product can be considered to utilize for heating column C-2.

5.8.4 Outlook on Hybrid Systems

The world is changing, and the chemical and biochemical industries have progressed to produce products in more sustainable ways (Lutze and Gorak, 2013). In particular, when changing from petroleum-based economy to bio-based,

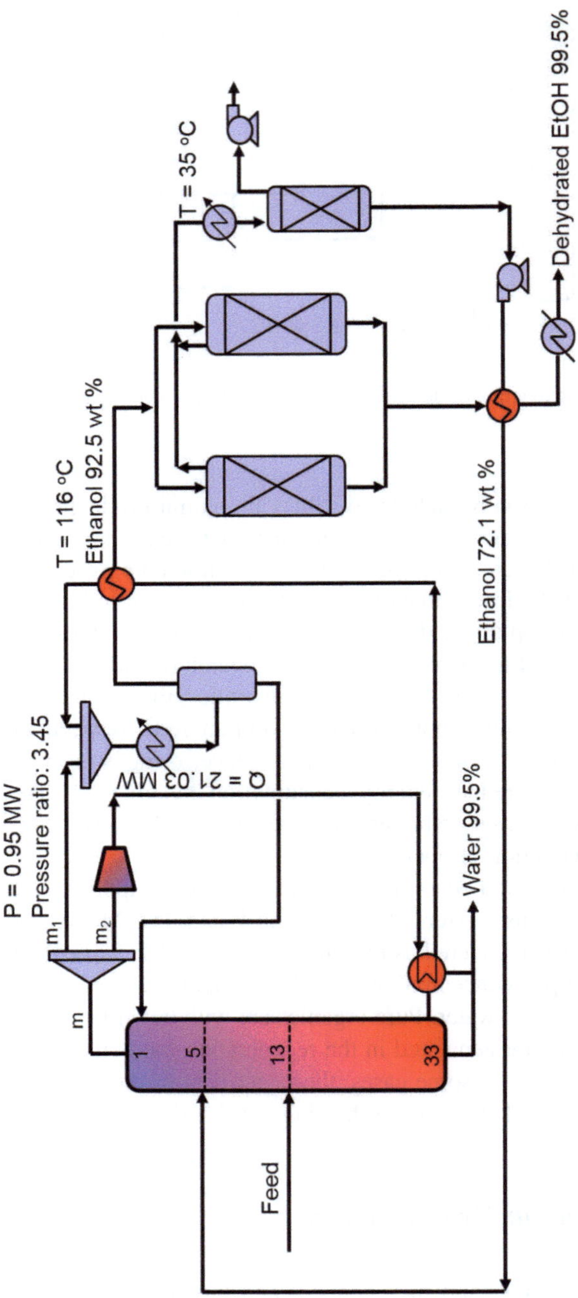

Fig. 5.44 Schematic diagram of the retrofitted hybrid distillation-adsorption system for CE production

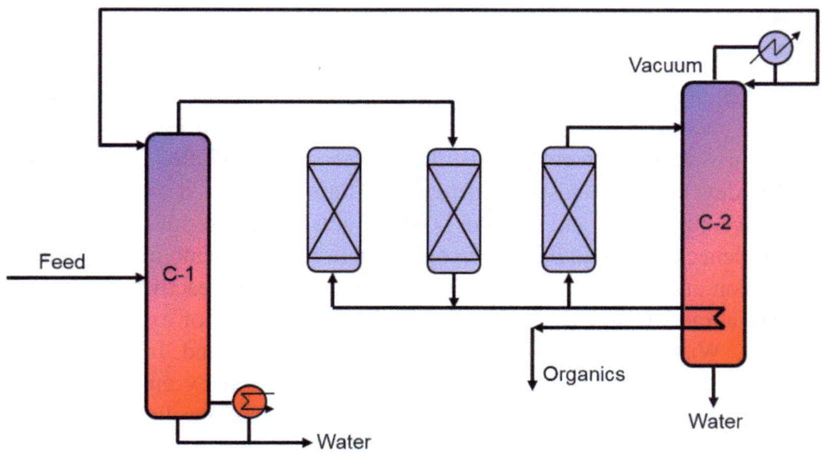

Fig. 5.45 Process for dewatering of organic compounds by distillation and adsorption (Mersmann et al., 2011)

distillation process still retains its significance as the main method for separating mixtures-although this old workhorse of the chemical industry is facing some new big and bold challenges (Kiss, 2013). Most biotechnological transformations lead to diluted systems, which need a wise hybrid system to reduce the role of distillation in product purification. Furthermore, highly integrated and intensified hybrid sequences, such as hybrid membrane/adsorption/distillation system or hybrid membrane/multi-effect RDWC system or hybrid reactive extraction/heat-pump-assisted reactive distillation system, are expected to break many challenges caused by emerging chemical and biochemical process industries in grass-roots and retrofit designs. The potential and feasibility of integration of different process intensification methods and different driving forces, such as microwaves, magnetism, gravity, and others in hybrid system to achieve synergy effects needs to be investigated and evaluated (Lutze and Gorak, 2013; Long et al., 2016).

To achieve the smooth development of hybrid systems and/or highly integrated and intensified hybrid sequences, future investigation will be necessary to consider the new catalysts, new stable and selective membrane materials, new environmentally friendly and selective solvents, new stable and selective adsorbent materials, cheap equipment fabrication, reliable control systems, and so on. Especially, the use of new enzymes, which are regioselective, within reactive distillation to enhance both reaction and separation has increased its research interest. Because enzymes are sensitive to higher temperatures, many works are needed to develop to find the common operating window between distillation and the enzymatic reaction (Lutze and Gorak, 2013; Lutze, 2014; Long et al., 2017). In addition, the development of thermodynamic data and models for description of integrated and intensified hybrid systems as well as development of synthesis/design tools are crucial to support the quick implementation of those hybrid systems.

5.9 Concluding Remarks

This chapter presented promising options for retrofit nonconventional distillation sequences, such as batch distillation, SSC, pressure swing azeotropic distillation, extractive distillation, azeotropic distillation, reactive distillation, and hybrid systems. The development and implementation efforts are focused on the applications of extractive, azeotropic and reactive DWC in grass-root cases as well as in retrofit and debottlenecking. In addition, a hybrid system using a membrane, or extraction, or adsorption and distillation, or TCDS, or DWC and even reactive separation processes are attractive options for improving the performance of existing distillation sequences. Wise hybrid systems or highly integrated and intensified hybrid sequences are expected to break many challenges caused by emerging chemical and biochemical process industries in both grass-roots and retrofit designs.

References

Adrian T, Bessling B, Hallmann H, Niekerken J (2000) DE 19860598
AEA Technology (2000) Reactive distillation.
Agrawal R, Fidkowski ZT (1999) New thermally coupled schemes for ternary distillation. AIChE J. 45:485–496
Agreda VH, Partin LR, Heise WH (1990) High-purity methyl acetate via reactive distillation. Chem Eng Prog. 86:40–46
Almeida-Rivera CP, Swinkels PLJ, Grievink J (2004) Designing reactive distillation processes: present and future. Comput Chem Eng. 28:1997–2020
Altman E, Stefanidis GD, Gerven Tv, Stankiewicz AI (2010) Process intensification of reactive distillation for the synthesis of n-propyl propionate: the effects of microwave radiation on molecular separation and esterification reaction. Ind Eng Chem Res. 49:10287–10296
Amminudin KA, Smith R (2001) Design and optimization of fully thermally coupled distillation columns. Part 2: application of dividing wall columns in retrofit. Chem Eng Res Des. 79:716–724
Babi DK, Gani R (2014) Hybrid distillation schemes: design, analysis, and application. In: Gorak A, Sorensen E (eds) Distillation: fundamentals and principles, 1st edn. Elsevier, New York, pp 225–270.
Behrens M, Olujic Z, Jansens PJ (2006) Combining reaction with distillation: hydrodynamic and mass transfer performance of modular catalytic structured packings. Chem Eng Res Des 84:381–389
Beneke, Peters M, Glasser D et al (2013) Understanding distillation using column profile maps. Wiley, Hoboken
Brambilla A (2014) Distillation control and optimization: operation fundamentals through software control. McGraw-Hill Professional, New York
Bravo-Bravo C, Segovia-Hernandez J, Gutierrez-Antonio C et al (2010) Extractive dividing wall column: design and optimization. Ind Eng Chem Res. 49:3672–3688
Buchaly C, Kreis P, Górak A (2007) Hybrid separation processes—combination of reactive distillation with membrane separation. Chem Eng Process. 46:790–799
Carmo MJ, Gubulin JC (1997) Ethanol-water adsorption on commercial 3A zeolites: kinetic and thermodynamic data. Braz J Chem Eng. 14:1–10
Cheng K, Wang SJ, Wong DSH (2013) Steady-state design of thermally coupled reactive distillation process for the synthesis of diphenyl carbonate. Comput Chem Eng. 52:262–271
DeGarmo JL, Parulekar VN, Pinjala V (1992) Consider reactive distillation. Chem Eng Prog. 88 (3):43

Demesa AG, Laari A, Tirronen E et al (2015) Comparison of solvents for the recovery of low-molecular carboxylic acids and furfural from aqueous solutions. Chem Eng Res Des. 93: 531–540

Doherty MF, Malone MF (2001) Conceptual design of distillation systems. Chemical engineering series. McGraw-Hill, New York

Drioli E, Brunetti A, Profio GD, Barbieri G (2012) Process intensification strategies and membrane engineering. Green Chem. 14:1561–1572

Drioli E, Curcio E (2007) Membrane engineering for process intensification: a perspective. J Chem Technol Biotechnol. 82:223–227

Emtir M, Etoumi A (2009) Enhancement of conventional distillation configurations for ternary mixtures separation. Clean Techn Environ Policy 11:123–131

Ennenbach F, Kolbe B, Ranke U (2000) Divided wall columns - a novel distillation concept. Petrol Techno Quart 5:97–103

Fidkowski Z, Krolikowski L (1986) Thermally coupled system of distillation columns: optimization procedure. AIChE J. 32:537–546

Fidkowski Z, Krolikowski L (1987) Minimum energy requirements of thermally coupled distillation systems. AIChE J. 33:643–654

Fontalvo J, Cuellar P, Timmer JMK et al (2005) Comparing pervaporation and vapor permeation hybrid distillation processes. Ind Eng Chem Res. 44:5259–5266

Fontalvo J, Keurentjes JTF (2015) A hybrid distillation–pervaporation system in a single unit for breaking distillation boundaries in multicomponent mixtures. Chem Eng Res Des. 99:158–164

Gebreslassie BH, Diwekar UM (2015) Efficient ant colony optimization for computer aided molecular design: case study solvent selection problem. Comput Chem Eng. 78:1–9

Gelbein AP, Buchholz M (1991) EP 428265A1

Genduso G, Amelio A, Colombini E, Luis P, Degreve J, Van der Bruggen B (2016) Retrofitting of extractive distillation columns with high flux, low separation factor membranes: A way to reduce the energy demand? Chem Eng Res Des. 109:127–140

Ghosh TK, Lin H-D, Hines AL (1993) Hybrid adsorption-distillation process for separating propane and propylene. Ind Eng Chem Res. 32:2390–2399

Gil ID, Botja DC, Ortiz P et al (2009) Extractive distillation of acetone/methanol mixture using water as entrainer. Ing Eng Chem Res. 48:4858–4865

Glinos KN, Malone MF (1985) Design of sidestream distillation columns. Ind Eng Chem Process Des Dev. 24:822–828

Glinos K, Malone MF (1988) Optimality regions for complex column alternatives in distillation systems. Chem Eng Res Des. 66:229–240

Glinos KN, Nikolaides IP, Malone MF (1986) New complex column arrangements for ideal distillation. Ind Eng Chem Process Des. Dev. 25:694–699

Gorak A, Stankiewicz A (2011) Intensified reaction and separation systems. Annu Rev Chem Biomol Eng. 2:431–451

Guedes BP, Feitosa MF, Vasconcelos LS et al (2007) Sensitivity and dynamic behavior analysis of an industrial azeotropic distillation column. Sep Purif Technol. 56:270–277

Gutiérrez-Guerra R, Segovia-Hernández JG, Hernández S (2009) Reducing energy consumption and CO_2 emissions in extractive distillation. Chem Eng Res Des 87:145–152

Harmsen GJ (2007) Reactive distillation: The front-runner of industrial process intensification: A full review of commercial applications, research, scale-up, design and operation. Chem Eng Process. 46:774–780

Harper PM, Gani R, Kolar P et al (1999) Computer-aided molecular design with combined molecular modeling and group contribution. Fluid Phase Equilib 158–160:337–347

Harwardt A, Marquardt W (2012) Heat-integrated distillation columns: vapor recompression or internal heat integration? AIChE J. 58:3740–3750

Holtbruegge J, Heile S, Lutze P et al (2013a) Synthesis of dimethyl carbonate and propylene glycol in a pilot-scale reactive distillation column: experimental investigation, modeling and process analysis. Chem Eng J. 234:448–463

Holtbruegge J, Wierschem M, Steinruecken S et al (2013b) Experimental investigation, modeling and scale-up of hydrophilic vapor permeation membranes: separation of azeotropic dimethyl carbonate/methanol mixtures. Sep Purif Technol. 118:862–878

Hostrup M, Harper PM, Gani R (1999) Design of environmentally benign processes: integration of solvent design and separation process synthesis. Comput Chem Eng. 23(10):1395–1414

Hsu K-Y, Hsiao Y-C, Chien I-L (2010) Design and control of dimethyl carbonate−methanol separation via extractive distillation in the dimethyl carbonate reactive-distillation process. Ind Eng Chem Res. 49(2):735–749

Huang K, Iwakabe K, Nakaiwa M et al (2005) Towards further internal heat integration in design of reactive distillation columns – Part I: the design principles. Chem Eng Sci. 4901–4914:60

Huang HJ, Ramaswamy S, Tschirner UW et al (2008) A review of separation technologies in current and future biorefineries. Sep Purif Technol. 62:1–21

Humbird D, Davis R, Tao L, et al (2011) Process design and economics for biochemical conversion of lignocellulosic biomass to ethanol: dilute–acid pretreatment and enzymatic hydrolysis of corn stover. Nnational renewable energy laboratory, Colorado

Hutchings GJ, Nicolaides CP (1992) Scurrell, developments in the production of methyl tert-butyl ether. Catal Today 15:23–49

Ibarra-Sánchez JdJ, Segovia-Hernández JG (2010) Reducing energy consumption and CO_2 emissions in extractive distillation: Part II. Dynamic behavior. Chem Eng Res Des. 88:135–145

Iglesias M, Orge B, Marino G et al (1999) Vapor−liquid equilibria for the ternary system acetone + methanol + water at 101.325 kPa. J Chem Eng Data 44:661–665

Jana K (2014) Advances in heat pump assisted distillation column: a review. Energ Convers Manage. 77:287–297

Jong WD, Marcotullio G (2010) Overview of biorefineries based on co-production of furfural, Existing concepts and novel developments. Int J Chem React Eng. 8:1–24

Karunanithi AT, Achenie LEK, Gani R (2005) A new decomposition−based computer−aided molecular/mixture design methodology for the design of optimal solvents and solvent mixtures. Ind Eng Chem Res. 44(13):4785–4797

Keller T (2014) Reactive distillation. In: Górak A, Olujic Z (eds) Distillation: equipment and processes, 1st edn. Elsevier, Oxford, pp 261–294

Kiss AA (2013) Advanced distillation technologies: design, control and applications, 1st edn. Wiley, West Sussex

Kiss AA, Bildea CS (2011) A control perspective on process intensification in dividing wall columns. Chem Eng Process. 50:281–292

Knapp JP, Doherty MF (1992) A new pressure-swing-distillation process for separating homogeneous azeotropic mixtures. Ind Eng Chem Res. 31:346–357

Kolbe B, Wenzel S (2004) Novel distillation concepts using one-shell columns. Chem Eng Process. 43:339–346

Kotai B, Lang P, Modla G (2007) Batch extractive distillation as a hybrid process: separation of minimum boiling azeotropes. Chem Eng Sci. 62:6816–6826

Kreis P, Górak A (2006) Process analysis of hybrid separation processes: combination of distillation and pervaporation. Chem EngRes Des 84:595–600

Lee HY, Jan CH, Chien IL et al (2010) Feed-splitting operating strategy of a reactive distillation column for energy-saving production of butyl propionate. J Taiwan Inst Chem Eng. 41:403–413

Lee MY, Minh LQ, Long NVD et al (2016) Retrofit of side stream columns to dividing wall columns, with a case study of industrial application, in gade pandu rangaiah. Chemical Process Retrofitting and Revamping Techniques and Applications. Wiley, West Sussex

Lee SK, Shin JH, Lee JK (2011) Divided wall distillation column for producing high purity normal butanol, and normal butanol distillation method. U.S. Patent US2011/0303526, 15 December

Lei Z, Chen B, Ding Z (2005) Special distillation processes. Elsevier, Amsterdam

Li M, Harten PF, Cabezas H (2002) Experiences in designing solvents for the environment. Ind Eng Chem Res. 41(23):5867–5877

Lipnizki F, Field RW, Ten PK (1999) Pervaporation-based hybrid process: a review of process design, applications andeconomics. J Membr Sci 153:183–210

Long NVD, Lee MY (2011) Improved energy efficiciency in debottlenecking using a fully thermally coupled distillation column. AsiaPac J Chem Eng. 6:338–348

Long NVD, Lee MY (2013a) Optimal retrofit of a side stream column to a dividing wall column for energy efficiency maximization. Chem Eng Res Des. 91:2291–2298

Long NVD, Lee MY (2013b) Optimal retrofit design of extractive distillation to energy efficient thermally couple distillation scheme. AIChE J. 59:1175–1182

Long NVD, Lee MY (2014) Review of retrofitting distillation columns using thermally coupled distillation sequences and dividing wall columns to improve energy efficiency. J Chem Eng Japan 47:87–108

Long NVD, Lee JT, Koo KK, Luis P, Lee MY (2017) Recent progress and novel applications in enzymatic conversion of carbon dioxide. Energies 10(4):473

Long NVD, Lee SH, Lee MY (2010) Design and optimization of a dividing wall column for debottlenecking of the acetic acid purification. Chem Eng Process. 49:825–835

Long NVD, Minh LQ, Ahmad F, Luis P, Lee MY (2016) Intensified distillation-based separation processes: recent developments and perspectives. Chem. Eng. Technol. 39:2183–2195

Long NVD, Minh LQ, Nhien LC, Lee MY (2015) A novel self-heat recuperative dividing wall column to maximize energy efficiency and column throughput in retrofitting and debottlenecking of a side stream column. Appl. Energy 159:28–38

Loy YY, Lee XL, Rangaiah GP (2015) Bioethanol recovery and purification using extractive dividing-wall column and pressure swing adsorption: an economic comparison after heat integration and optimization. Sep Purif Technol. 149:413–427

Luis P, Amelio A, Vreysen S, Calabro V, Van der Bruggen B (2014) Simulation and environmental evaluation of process design: Distillation vs. hybrid distillation–pervaporation for methanol/tetrahydrofuran separation. Appl. Engergy 113:565–575

Lutze P (2014) Distillation in bioprocessing. In: Gorak A, Schoenmakers H (eds) Distillation: operation and applications, 1st edn. Elsevier, Oxford

Lutze P, Dada EA, Gani R, Woodley JM (2010) Heterogeneous catalytic distillation - a patent review. Recent Pat Chem Eng. 3:208–229

Lutze P, Gorak A (2013) Reactive and membrane-assisted distillation: Recent developments and perspective. Chem Eng Res Des. 91:1978–1997

Luyben WL (2005) Comparison of pressure-swing and extractive-distillation methods for methanol-recovery systems in the TAME reactive-distillation process. Ind Eng Chem Res. 44:5715–5725

Luyben WL (2008a) Design and control of a fully heat-integrated pressure swing azeotropic distillation system. Ind Eng Chem Res. 47:2681–2695

Luyben WL (2008b) Comparison of extractive distillation and pressure-swing distillation for acetone-methanol separation. Ind. Eng. Chem. Res. 47:2696—2707

Luyben WL (2008c) Control of the maximum-boiling acetone/chloroform azeotropic distillation system. Ind Eng Chem Res. 47:6140–6149

Luyben WL, Chien IL (2010) Design and control of distillation systems for separating azeotropes. Wiley-AIChE, Hoboken

Luyben WL, Yu C-C (2008) Reactive distillation design and control. Wiley-AIChE, Hoboken

Macedonio F, Drioli E (2017) Membrane engineering for green process engineering. Engineering 3:290–298

Mersmann A, Kind M, Stichlmair J (2011) Thermal separation technology – principles, methods, process design. Springer, Verlag Berlin, Heidelberg

Modla G, Lang P (2013) Heat pump systems with mechanical compression for batch distillation. Energy 62:403–417

Moritz P, Hasse H (1999) Fluid dynamics in reactive distillation packing Katapak®-S. Chem Eng Sci. 54:1367–1374
Moulijn JA, Kreutzer MT, Nijhuis TA, Kapteijn F (2011) Monolithic catalysts and reactors: high precision with low energy consumption. Adv Catal 54:249–327
Moulijn JA, Stankiewicz A, Grievink J et al (2008) Process intensification and process systems engineering: a friendly symbiosis. Comput Chem Eng. 32:3–11
Nhien LC, Long NVD, Kim S et al (2016a) Design and optimization of intensified biorefinery process for furfural production through a systematic procedure. Biochem Eng J. 116:166–175
Nhien LC, Long NVD, Kim S et al (2016c) Design and assessment of hybrid purification processes through a systematic solvent screening for the production of levulinic acid from lignocellulosic biomass. Ind Eng Chem Res. 55:5180–5189
Nhien LC, Long NVD, Kim SY, Lee MY (2017a) Techno-economic assessment of hybrid extraction and distillation processes for furfural production from lignocellulosic biomass. Biotechnol Biofuels 10:81
Nhien LC, Long NVD, Lee MY (2016b) Design and optimization of the levulinic acid recovery process from lignocellulosic biomass. Chem Eng Res Des. 107:126–136
Nhien LC, Long NVD, Lee MY (2017b) Novel heat-integrated and intensified biorefinery process for cellulosic ethanol production from lignocellulosic biomass. Energ Convers Manage. 141:367–377
Niesbach A, Adams II TA, Lutze P (2013) Semicontinuous distillation of impurities for the production of butyl acrylate from bio-butanol and bioacrylic acid. Chem Eng Proc. 74:165–177
Nikolaides IP, Malone MF (1987) Approximate design of multiple-feed/side-stream distillation systems. Ind Eng Chem Res. 26:1839–1845
Papastathopoulou HS, Luyben WL (1991) Control of a binary sidestream distillation column. Ind Eng Chem Res. 30:713–721
Pyhälahti A, Klemola K (1998) Bench scale experiments with TAME for the comparison of two catalytic distillation packing arrangements. Acta Polytech Scand 257:101–111
Ritter JA, Wu F, Ebner AD (2012) New approach for modeling hybrid pressure swing adsorption–distillation processes. Ind Eng Chem Res. 51(27):9343–9355
Rooks RE, Malone MF, Doherty MF (1996) A geometric design method for side-stream distillation columns. Ind Eng Chem Res. 35:3653–3664
Roth T, Kreis P, Górak A (2013) Process analysis and optimisation of hybrid processes for the dehydration of ethanol. Chem Eng Res Des. 91:1171–1185
Shizou M, Zheng SN, Ikuho Y (2000) Analysis of divided-wall column for extractive distillation. Kagaku Kogaku Ronbunshu 26:627–632
Sirkar KK, Fane AG, Wang R, Wickramasinghe SR (2015) Process intensification with selected membrane processes. Chem Eng Process. 87:16–25
Skiborowski M, Gorak A (2016) Hybrid separation processes. In: Lutze P, Gorak A (eds) Reactive and membrane-assisted separations. De Gruyter, Berlin
Skiborowski M, Harwardt A, Marquardt W (2013) Conceptual design of distillation-based hybrid separation processes. Annu Rev Chem Biomol Eng. 4:45–68
Slade B, Stober B, Simpson D (2006) Dividing wall column revamp optimizes mixed xylenes production. IChemE Symp Ser. 152
Smith, LA (1980) Catalyst system for separating isobutene from C streams. U.S. patent 4,215,011, 27 July
Smith R (2005) Chemical process design and integration. Wiley, West Sussex, p 212–232
Sowerby B, Crittenden BD (1988) An experiment comparison of type A molecular sieves for drying the ethanol–water azeotrope. Gas Sep Purif. 2:77–83
Stephan W, Noble RD, Koval CA (1995) Design methodology for a membrane/distillation column hybrid process. J Membr Sci. 99(3):259–272
Stichlmair JG, Fair JR (1998) Distillation - principles and practice. Wiley–VCH, New York
Stringaro JP (1995) EP 631813

Subawalla H, Fair JR (1999) Design guidelines for solid-catalyzed reactive distillation systems. Ind Eng Chem Res. 38:3696–3709

Sundmacher K, Hoffmann U (1993) Activity evaluation of a catalytic distillation packing for MTBE production. Chem Eng Technol. 16:279

Sundmacher K, Kienle A (2003) Reactive distillation—status and future directions. Wiley–VCH, New York

Tijing LD, Woo YC, Choi JS, Lee SH, Kim SH, Shon HK (2015) Fouling and its control in membrane distillation - A review. J Membr Sci. 475:215–244

Treybal RE (1955) Mass-transfer operations. McGraw Hill, New York

Van Hoof V, Van den Abeele L, Buekenhoudt A, Dotremont C, Leysen R (2004) Economic comparison between azeotropic distillation and different hybrid systems combining distillation with pervaporation for the dehydration of isopropanol. Separation Purif Technol. 37:33–49

Venkatesan S (2013) Adsorption. In: Ramaswamy S, Huang HJ, Ramarao BV (eds) Separation and purification technologies in biorefineries. Wiley, West Sussex

Vercher E, Orchilles AV, Miguel PJ et al (2006) Isobaric vapor-liquid equilibria for acetone + methanol + lithium nitrate at 100 kPa. Fluid Phase Equilib. 250:131–137

Verhoef A, Degrève J, Huybrechs B et al (2008) Simulation of a hybrid pervaporation–distillation process. Com Chem Eng. 32:1135–1146

Verhoeve L, De Schepper H (1973) The vapor - liquid equilibria of the binary, ternary and quaternary systems formed by acetone, methanol, propan-2-ol and water. J Appl Chem Biotechnol. 23:607–619

Wachter JA, Ko TKT, Andres RP (1988) Minimum reflux behavior of complex distillation columns. AIChE J. 34:1164–1184

Wang SJ, Yu CC, Huang HP (2010) Plant-wide design and control of DMC synthesis process via reactive distillation and thermally coupled extractive distillation. Com Chem Eng. 34:361–373

Wankat PC (2012) Separation process engineering. Includes mass transfer analysis, 3rd edn. Prentice-Hall, Upper Saddle River,

Werth K, Lutze P, Kiss AA, Stankiewicz AI, Stefanidis GD, Górak A (2015) A systematic investigation of microwave-assisted reactive distillation: Influence of microwaves on separation and reaction. Chem Eng Process. 93:87–97

Widagdo S, Seider WD (1996) Azeotropic distillation. AIChE J. 42:96–130

Yu B, Wang Q, Xu C (2012) Design and control of distillation system for methylal/methanol separation. Part 2: pressure swing distillation with full heat integration. Ind Eng Chem Res. 51:1293–1310

Epilogue

This book has compiled a set of practical solutions for distillation retrofit in the wake of concerns on increase of fossil energy price, capacity as well as on environmental regulation. The concepts considered in this book help practical distillation engineers and researchers to develop a solution fast and efficiently for improving energy efficiency and/or capacity of distillation sequence. By adding to or replacing the whole or part of one type of equipment with a better alternative, and reorganizing the process involving several process steps, target key goals, such as conserving natural resources, reducing energy required and/or operating cost, improving conversion and/or selectivity of reactions, enhancing productivity, using feed of different quality and/or alternative feed, achieving the new specifications of products, enhancing the control of the process, improving the safety, reliability and flexibility of the process and mitigating the discharge of ecologically harmful materials, can be achieved.

Modifying or replacing existing column internals can not only improve separation efficiency, but also govern the column hydraulic performance. It becomes an attractive option for many retrofit projects. Thus nowadays many types of high performance trays/packings have been produced and are expected to improve the capacity and energy efficiency of distillation. However, to achieve successful project, engineers should be aware of hydraulic and mass transfer aspects of the tray operation such as entrainment and downcomer flooding.

Recent studies and industrial applications have proven that advanced process integration techniques such as multi-effect distillation and heat pump assisted distillation, process intensification techniques, such as TCDS, DWC and reactive distillation, and innovative hybrid systems, such as hybrid membrane-RDWC configuration, could reduce the required energy for the separation of mixtures into pure product streams. These schemes make significant contributions to improving the energy efficiency and/or capacity and mitigating CO_2 in the retrofit of distillation processes. They are becoming a dominant trend in the chemical process industries. Note that the design of a retrofit must attempt to work within the constraints of existing equipment. Thus, the current operating conditions, maximum

N.V.D. Long, M. Lee, *Advances in Distillation Retrofit*, DOI 10.1007/978-981-10-5901-8

temperature, decomposition, fouling tendencies, construction material, lifetime, component characteristics, as well as available area for the new installed equipment and so on must be carefully considered and evaluated to determine the best retrofit option.

In biochemical and biotechnology processes, the unit operations for the separation of mixture including biological molecules differ from their counterparts in the chemical industry. Much effort needs to investigate and evaluate in this area when applying process integration and process intensification techniques to retrofit distillation processes. Especially, the use of enzymes within reactive separation to enhance both reaction and separation has increased its research interest. Because enzymes are sensitive to higher temperatures, many works are needed to develop. In addition, the development of thermodynamic data and models for the description of systems containing enzyme as well as the development of synthesis/design tools are essential for quick implementation of those systems. Most biotechnological transformations lead to diluted systems, which need a wise hybrid system to reduce the role of distillation in product purification.

Appendix

a. Sizing the column: The column diameter was determined under a column flooding condition that fixed the upper limit of the vapor velocity. The operating velocity is normally 70–90% of the flooding velocity (Sinnott, 2005).
b. Capital cost: Guthrie's modular method was applied (Biegler et al., 1997). The investment cost for conventional distillation is the total cost of the column and auxiliary equipment, such as reboilers and condensers. For the DWC, it includes the additional dividing wall cost.

$$\text{Tray stack} = (N-1) \times \text{tray spacing} \tag{A.1}$$

$$\text{Total height} = \text{Tray stack} + \text{Extra feedspace} + \text{Disengagement} + \text{Skirt height} \tag{A.2}$$

$$\text{Updated bare module cost(BMC)} = \text{UF} \times \text{BC} \times (\text{MPF} + \text{MF} - 1) \tag{A.3}$$

$$\text{where UF is the update factor: } UF = \frac{\textit{present cost index}}{\textit{base cost index}} \tag{A.4}$$

$$\text{BC is bare cost, for vessels: } BC = BC_0 \times \left(\frac{L}{L_0}\right)^{\alpha} \times \left(\frac{D}{D_0}\right)^{\beta} \tag{A.5}$$

$$\text{for the heat exchanger: } BC = BC_0 \times \left(\frac{A}{A_0}\right)^{\alpha} \tag{A.6}$$

$$\text{Area of heat exchanger}, A = \frac{Q}{U\Delta T} \tag{A.7}$$

$$\text{For the compressor: } BC = BC_0 \times \left(\frac{S}{S_0}\right)^{\alpha} \tag{A.8}$$

N.V.D. Long, M. Lee, *Advances in Distillation Retrofit*, DOI 10.1007/978-981-10-5901-8

where MPF is the material and pressure factor; MF is the module factor (typical value), which is affected by the base cost. *D*, *L* and *A* are the diameter, length and area, respectively. S is the brake horsepower.

$$\text{Updated bare module cost for tray stack(BMC)} = \text{UF} \times \text{BC} \times \text{MPF} \quad \text{(A.9)}$$

$$\text{The material and pressure factor: MPF} = \text{F}_\text{m} + \text{F}_\text{s} + \text{F}_\text{m} \quad \text{(A.10)}$$

c. Operating cost (Op):

$$\text{Op} = \text{C}_{\text{steam}} + \text{C}_{\text{CW}} + \text{C}_{\text{electricity}} + \text{C}_{\text{refrigeration}} \quad \text{(A.11)}$$

where C_{steam} is the cost of the steam; C_{CW} is the cost of cooling water; $C_{electricity}$ is the cost of electricity; and $C_{refrigeration}$ is the cost of refrigeration

d. $$\text{Cost saving} = \text{Operating cost saving} - \text{modification cost} \quad \text{(A.12)}$$

e. $$\text{Payback period} = \text{cost of project/saving per year} \quad \text{(A.13)}$$

f. Estimation of the CO_2 emission (Gadalla et al., 2005)

In the combustion of fuels, air is assumed to be in excess to ensure complete combustion, so that no carbon monoxide is formed. The amount of CO_2 emitted, $[CO_2]_{Emissions}$ (kg/s), is related to the amount of fuel burnt, Q_{Fuel} (kW), in the heating device, as follows:

$$[CO_2]_{\text{Emissions}} = \left(\frac{Q_{\text{Fuel}}}{\text{NHV}}\right)\left(\frac{C\%}{100}\right)\alpha \quad \text{(A.14)}$$

where α (=3.67) is the ratio of the molar masses of CO_2 and C, while NHV, which is equal to 47141 (kJ/kg), represents the net heating value of natural gas with a carbon content of 75%.

The amount of fuel burnt can be calculated using the following equation:

$$Q_{\text{Fuel}} = \frac{Q_{\text{Proc}}}{\lambda_{\text{Proc}}}(h_{\text{Proc}} - 419)\frac{T_{\text{FTB}} - T_\text{o}}{T_{\text{FTB}} - T_{\text{stack}}} \quad \text{(A.15)}$$

where λ_{Proc} (kJ/kg) and h_{Proc} (kJ/kg) are the latent heat and enthalpy of steam delivered to the process, respectively, while T_{FTB} (°C) and T_{stack} (°C) are the flame and stack temperatures, respectively. The flame temperature of 1800 °C and stack temperature of 160 °C are used in the calculations.

References

Biegler LT, Grossmann IE, Westerberg AW (1997) Systematic methods of chemical process design. Prentice Hall Inc., pp 110–141

Gadalla MA, Olujic Z, Jansens PJ, et al. (2005) Reducing CO_2 emissions and energy consumption of heat-integrated distillation systems. Environ Sci Technol. 39: 6860–6870

Sinnott SK (2005) Chemical engineering design (4th edn). In: Coulson, Richardson (eds) Chemical engineering series (vol. 6). Elsevier Butterworth Heinemann, Oxford

Index

N.V.D. Long, M. Lee, *Advances in Distillation Retrofit*, DOI 10.1007/978-981-10-5901-8

H

I

K

L

M

MIX
Papier aus verantwortungsvollen Quellen
Paper from responsible sources
FSC® C105338

If you have any concerns about our products,
you can contact us on
ProductSafety@springernature.com

In case Publisher is established outside the EU,
the EU authorized representative is:
Springer Nature Customer Service Center GmbH
Europaplatz 3, 69115 Heidelberg, Germany

Printed by Libri Plureos GmbH
in Hamburg, Germany